On Fact and Fraud

⌐ On Fact and Fraud

Cautionary Tales

from the Front Lines

of Science

David Goodstein

PRINCETON UNIVERSITY PRESS

Princeton and Oxford

Copyright © 2010 by Princeton University Press

Published by Princeton University Press, 41 William Street,

Princeton, New Jersey 08540

In the United Kingdom: Princeton University Press, 6 Oxford Street,

Woodstock, Oxfordshire OX20 1TW

All Rights Reserved

Library of Congress Cataloging-in-Publication Data

Goodstein, David L., 1939–

  On fact and fraud : cautionary tales from the front lines of science / David
Goodstein.

     p. cm.

  Includes bibliographical references and index.

  ISBN 978-0-691-13966-1 (cloth : alk. paper)

  1. Fraud in science. 2. Research—Moral and ethical aspects. I. Title.

  Q175.37.G66 2010

  500—dc22   2009039252

British Library Cataloging-in-Publication Data is available

This book has been composed in Arno Pro with DIN Pro display

Printed on acid-free paper. ∞

press.princeton.edu

Printed in the United States of America

10 9 8 7 6 5 4 3 2 1

This book is dedicated

to my two wonderful children,

Marcia and Mark,

and to their spouses,

Bill and Brence.

⊓ Contents

⟲ Illustrations

⌐ Preface

This book is, in a sense, the culmination of a lifetime spent in science and in science administration at the California Institute of Technology, where I've been a professor of physics and applied physics for more than forty years. In 1988, I became Caltech's vice provost, and soon after I settled into my new office I found myself in charge of all cases of scientific misconduct, real or imagined, that arose at the institute. After a number of years in this arcane field, I decided to avail myself of one the great privileges that comes with being a professor—the opportunity to share new knowledge—and I proposed, along with my colleague Jim Woodward, a professor of philosophy, to teach a course in scientific fraud. At least that's what we wanted to call it, but the institute's Faculty Board, in its wisdom, didn't want us teaching anything with that title to the students. So we wound up calling our new course "Scientific Ethics" and taught it annually for the next ten years.

When I stepped down from the vice provost's position in 2007, I realized that I now had the time to acquaint a much larger audience with these issues, by writing a book. Regardless of whether we call our subject fraud or ethics, this book will be a series of personal reflections on the topic, focusing on cases in which I have been involved during my career. Some of these are likely to be new to readers, while others will be familiar but enlivened, I hope, by the introduction of new material.

Similarly, I will have relatively little to say about some famous instances of alleged scientific misconduct, although two of the best known—the Baltimore and Gallo cases (which have been extensively documented and described by other authors)—are mentioned when I talk about federal efforts to get a handle on the whole science fraud issue in the 1990s.

During the years that Jim and I taught our science ethics course, we often found ourselves lamenting the lack of a suitable textbook. In particular, we always felt that such a course must be based on real case histories, not made-up ones. Finding such material in those instances where the accused have been publicly exonerated has not been difficult, but it has been a different story where fraud has actually been committed, because for some reason people in this field have often found it necessary to impose confidentiality to protect the guilty. Thus a lively cottage industry has grown up around presenting fabricated case histories that showcase various ethical dilemmas. However valuable and well intentioned such scenarios are, they cannot possess the immediacy of real cases, and so this book, which does present real cases, may yet serve as a textbook on the subject.

The book opens with a look at the subject of fraud in science within the larger context of how real scientists operate in the real world. In particular, I specify some fifteen seemingly plausible ethical principles for science and then proceed to demolish them one by one as realistic guides to sound scientific conduct. The chapter concludes with a kind of user's manual on how to succeed in science without ever raising the specter of fraud.

The bulk of the book consists of the actual cases. First up is the matter of Nobel laureate Robert A. Millikan, who has been accused of misconduct in his determination of the charge on the electron. It's a topic that has generated lively debate and has more than a passing interest for me, given that Millikan was the founding president of the university where I have spent my entire professional life. It's a close call, but ultimately the verdict is not guilty. I then fast-forward to modern-day Caltech and examine two instances of misconduct that occurred in the lab of one of the institute's leading biologists in the 1990s. This was also the decade in which the United States government first ventured onto the treacherous terrain of regulating ethics in science, encountering the usual hazards one finds in such minefields, and I relate how both federal officials and my own institution, Caltech, developed oversight protocols to deal with scientific fraud (Caltech's *Policy on Research Misconduct*, which I drew up in my capacity as vice provost, is included as an appendix to this volume). Next up is the case of cold fusion, where we find confusion and controversy of the most fascinating kind, but no clear evidence of fraud. Finally, I look at two instances of misconduct in physics and consider the discovery of high-temperature superconductivity, an illuminating and perhaps unique instance in which the impossible actually occurred in an utterly irreproachable manner.

The net result is something of a standoff: four cases of fraud, two cases where no fraud happened, and a case where no fraud occurred but a phenomenon that flew in the face of nearly everything we thought we knew did. Science is a wonderful enterprise. We are forever learning new things about the universe we inhabit. The great majority of scientists are honorable people who will fiercely protect the validity of the science they do. Nevertheless, every once in a while, along comes someone who would undermine the enterprise. We must be vigilant to find and expose such wrongdoers, careful at the same time not to spread the blame beyond where it belongs and unintentionally stifle the freedom to question and explore that has always characterized scientific progress. I hope the reader will take this book in that spirit.

On Fact and Fraud

⌐ One
Setting the Stage

Fraud in science is, in essence, a violation of the scientific method. It is feared and denigrated by all scientists. Let's look at a few real cases that have come up in the past.

Piltdown Man, a human cranium and ape jaw found in a gravel pit in England around 1910, is perhaps the most famous case. Initially hailed as the authentic remnants of one of our more distant ancestors, the interspecies skeletal remains were exposed as a fraud by modern dating methods in 1954. To this day no one knows who perpetrated the deception or why. One popular theory is that the perpetrator was only trying to help along what was thought to be the truth. Prehistoric hominid remains had been discovered in France and Germany, and there were even rumors of findings in Africa. Surely humanity could not have originated in those uncivilized places. Better to have human life begin in good old England!

As it turned out, the artifact was rejected by the body of scientific knowledge long before modern dating methods showed it to be a hoax. Growing evidence that our ancient forebears looked nothing like Piltdown Man made the discovery an embarrassment at the fringes of anthropology. The application of modern dating methods confirmed that both artifacts were not much older than their discovery date.

Sir Cyril Burt was a famous British psychologist who studied the heritability of intelligence by means of identical twins who had been separated at birth. Unfortunately there seem not to have been enough such convenient subjects to study, so he apparently invented thirty-three additional pairs, and because that gave him more work than he could handle, he also invented two assistants to take care of them. His duplicity was uncovered in 1974, some three years after his death.

That same year, William Summerlin, a researcher at the Sloan-Kettering Institute for Cancer Research in New York City, conducted a series of experiments aimed at inducing healthy black skin grafts to grow on a white mouse. Evidently, nature wasn't sufficiently cooperative, for he was caught red-handed trying to help her out with a black felt-tipped pen.

John Darsee was a prodigious young researcher at Harvard Medical School, turning out a research paper about once every eight days. That lasted a couple of years until 1981, when he was caught fabricating data out of whole cloth.

Stephen Breuning was a psychologist at the University of Pittsburgh studying the effects of drugs such as Ritalin on patients. In 1987 it was determined that he had fabricated data. His case was particularly bad, because protocols for treating patients had been based on his spurious results.

Science is self-correcting, in the sense that a falsehood injected into the body of scientific knowledge will eventually be discovered and rejected. But that fact does not protect the scientific enterprise against fraud, because injecting falsehoods into the body of science is rarely, if ever, the purpose of those who perpetrate fraud. They almost always believe that they are injecting a truth into the scientific record, as in the cases above, but without going through all the trouble that the real scientific method demands.

That's why science needs active measures to protect it. Fraud, or misconduct, means dishonest professional behavior, characterized by the intent to deceive—the very antithesis of ethical behavior in science. When you read a scientific paper, you are free to agree or disagree with its conclusions, but you must always be confident that you can trust its account of the procedures that were used and the results produced by those procedures.

For years it was thought that scientific fraud was almost always restricted to biomedicine and closely related sciences, and although there are exceptions, most instances do surface in these fields. There are undoubtedly many reasons for this curious state of affairs. For example, many misconduct cases involve medical doctors rather than scientists with Ph.D.s (who are trained to do research). To a doctor, the welfare of his or her patient may be more important than scientific truth. In a case that came up in the 1980s, for example, a physician in Montreal was found to have falsified the records of participants in a large-scale breast-cancer study. Asked why he did it, he said it was in order to get better medical care for his patients. However, the greater number of cases arises from more self-interested motives. Although the perpetrators usually think that they're doing the right thing, they also know that they're committing fraud.

In recent cases of scientific fraud, three motives, or risk factors, have always been present. In nearly all cases, the perpetrators

1. were under career pressure;
2. knew, or thought they knew, what the answer to the problem they were considering would turn out to be if they went to all the trouble of doing the work properly; and

3. were working in a field where individual experiments are not expected to be precisely reproducible.

It is by no means true that fraud always arises when these three factors are present. In fact, just the opposite is true: These factors are often present, and fraud is quite rare. But they do seem to be present whenever fraud occurs. Let us consider them one at a time.

Career pressure. This is clearly a motivating factor, but it does not offer us any special insights into why a small number of scientists stray professionally when most do not. All scientists, at all levels, from fame to obscurity, are pretty much always under career pressure. On the other hand, simple monetary gain is seldom if ever a factor in scientific fraud.

Knowing the answer. Scientific fraud is almost always a transgression against the methods of science, not purposely against the body of knowledge. Perpetrators think they know how the experiment would come out if it were done properly, and they decide that it is not necessary to go to all the trouble of doing it properly.

Reproducibility. In reality, experiments are seldom repeated by others in science. Nevertheless, the belief that someone else can repeat an experiment and get—or not—the same result can be a powerful deterrent to cheating. Here a pertinent distinction arises between biology and the other sciences, in that biological variability may provide apparent cover for a biologist who is tempted to cheat. Sufficient variability exists among organisms that the same procedure, performed on two test subjects as nearly identical as possible, is not expected to give exactly the same result. If two virtually identical rats are treated with the same carcinogen, they are not expected to develop the same tumor in the same place at the same time. This last point certainly helps to explain why scientific fraud is found mainly in the biomedical area. (Two cases in physics offer an interesting test of this hypothesis. They are addressed in more detail later in this volume.)

No human activity can stand up to the glare of relentless, absolute honesty. We build little hypocrisies and misrepresentations into what we do to make our lives a little easier, and science, a very human enterprise, is no exception. For example, every scientific paper is written as if the particular investigation it describes were a triumphant progression from one truth to the next. All scientists who perform research, however, know that every scientific experiment is chaotic—like war. You never know what's going on; you cannot usually understand what the data mean. But in the end you figure out what it was all about, and then, with hindsight, you write it up as one clear and certain step after another. This is a kind of hypocrisy, but one that is deeply embedded in the way we do science. We are so accustomed to it that we don't even regard it as a misrepresentation. Courses are not offered in the rules of misrepresentation in scientific papers, but the apprenticeship that one goes through to become a scientist does involve learning them.

The same apprenticeship, however, also inculcates a deep respect for the inviolability of scientific data and instructs the neophyte scientist in the ironclad distinction between harmless fudging and real fraud. For example, it may be marginally acceptable, in writing up your experiment, to present your best data and casually refer to them as typical (because you mean typical of the phenomenon, not typical of your data), but it is not acceptable to move one data point just a little bit to make the data look better. All scientists would agree that to do so is fraud. That is because experiments must deal with physical reality, a major point that can only be assured by an honest presentation of all the data.

In order to define as precisely as possible what constitutes scientific misconduct or fraud, we need first to have the clearest possible understanding of how science actually works. Otherwise, it is all too easy to formulate plausible-sounding ethical principles that would be unworkable or even damaging to the scientific enterprise if they were actually put into practice. Here, for example, is a plausible but unworkable set of such precepts.

1. A scientist should never be motivated to do science for personal gain, advancement, or other rewards.
2. Scientists should always be objective and impartial when gathering data.
3. Every observation or experiment must be designed to falsify a hypothesis.
4. When an experiment or an observation gives a result contrary to the prediction of a certain theory, all ethical scientists must abandon that theory.
5. Scientists must never believe dogmatically in an idea or use rhetorical exaggeration in promoting it.
6. Scientists must "bend over backwards" (in the words of iconic physicist Richard Feynman)[1] to point out evidence that is contrary to their own hypothesis or that might weaken acceptance of their experimental results.

7. Conduct that seriously departs from commonly accepted behavior in the scientific community is unethical.

8. Scientists must report what they have done so fully that any other scientist can reproduce the experiment or calculation. Science must be an open book, not an acquired skill.

9. Scientists should never permit their judgments to be affected by authority. For example, the reputation of the scientist making a given claim is irrelevant to the validity of the claim.

10. Each author of a multi-author paper is responsible for every part of the paper.

11. The choice and order of authors on a multi-author paper must strictly reflect the contributions of the authors to the work in question.

12. Financial support for doing science and access to scientific facilities should be shared democratically, not concentrated in the hands of a favored few.

13. There can never be too many scientists in the world.

14. No misleading or deceptive statement should ever appear in a scientific paper.

15. Decisions about the distribution of scientific resources and publication of experimental results must be guided by the judgment of scientific peers who are protected by anonymity.

Let's now look at each of our *diktats* in turn, beginning with principle 1. In a parallel case in economic life, well-intentioned attempts to eliminate the role of greed or speculation can have disastrous consequences. In fact, seemingly bad behavior such as the aggressive pursuit of self-interest can, in a properly function-ing system, produce results that are generally beneficial.

Principles 2 and 3 derive from the following arguments. According to Francis Bacon, who set down these ideas in the seventeenth century, science begins with the careful record-ing of observations.[2] These should be, insofar as is humanly possible,uninfluenced by any prior prejudice or theoretical preconception. When a large enough body of observations is present, one generalizes from these to a theory or hypothesis by a process of induction—that is, working from the specific to the general.

Historians, philosophers, and those scientists willing to venture into such philosophic waters are virtually unanimous in rejecting Baconian inductivism as a general characterization of

Figure 1.1
Engraved portrait of English philosopher and essayist Sir Fran-cis Bacon, by Dutch engraver Jacobus Houbraken (1698–1780), Amsterdam, dated 1738, possibly after a portrait painting done circa 1731 by John Vanderbank (1694–1735). Courtesy of California Institute of Technology Archives.

good scientific method (adieu, principle 2). You cannot record all that you observe; some principle of relevance is required. But decisions about what is relevant depend on background assumptions that are highly theoretical. This is sometimes expressed by saying that all observation in science is "theory-laden" and that a "theoretically neutral" language for recording observations is impossible.

The idea that science proceeds only and always by means of inductive generalization is also misguided. Theories in many parts of science have to do with things that can't be directly observed at all: forces, fields, subatomic particles, proteins, and so on. For this and many other reasons, no one has been able to formulate a defensible theory of Baconian inductivist science. Although few scientists believe in inductivism, many have been influenced by the falsifiability ideas of the twentieth-century philosopher Karl Popper.[3] According to these ideas, we assess the validity of a hypothesis by extracting from it a testable prediction. If the test proves the prediction to be false, the hypothesis is also by definition false and must be rejected. The key point to appreciate here is that no matter how many observations agree with the prediction, they will never suffice to prove that the prediction is true, or verified, or even more probable than it was before. The most that we are allowed to say is that the theory has been tested and not yet falsified. Thus an important asymmetry informs the Popperian model of verification and falsification. We can show conclusively that a hypothesis is false, but we can never demonstrate conclusively that it is true. In this view, science proceeds entirely by showing that seemingly sound ideas are wrong, so that they must be replaced by better ideas.

Inductivists place much emphasis on avoidance of error. By contrast, falsifiability advocates believe that no theory can ultimately be proved right, so our aim should be to detect errors and learn from them as efficiently as possible. Thus, a laudable corollary of the Popperian view is that if science is to progress, scientists must be free to be wrong.

But falsifiability also has serious deficiencies. Testing a given hypothesis, H, involves deriving from it some observable consequence, O. But in practice, O may depend on other assumptions, A (auxiliary assumptions, philosophers call them). So if H is false, it may be that O is false, but it may also be that O is true and A is false.

One immediate consequence of this simple logical fact is that the asymmetry between falsifiability and verification vanishes. We may not be able to conclusively verify a hypothesis, but we can't falsify it either. Thus it may be a good strategy to hang onto a hypothesis even when an observation seems to imply that it's false. The history of science is full of examples of this sort of anti-Popperian strategy succeeding where a purely Popperian strategy would have failed. Perhaps the classic example is Albert Einstein's seemingly absurd conjecture that the speed of light must be the same for all observers, regardless of their own speed. Many observations had shown that the apparent speed of an object depends on the speed of the observer. But those observations were not true for light, and the result was the special theory of relativity (so much for principle 3).

Both inductivism and falsifiability envision the scientist encountering nature all alone. But science is carried out by a community of investigators. Suppose a scientist who has devoted a great deal of time and energy developing a theory is faced with a decision about whether to hold onto it in the face of some contrary evidence. Good Popperian behavior would be to give it up, but the communal nature of science suggests another possibility. Suppose our scientist has a rival who has invested time and energy developing an alternative theory. Then we can expect the rival to act as a severe Popperian critic of the theory. As long as others are willing to do the job, our scientist need not take on the psychologically daunting task of playing his own devil's advocate. In fact, scientists, like other people, find it difficult to commit to an arduous long-term project if they spend too much time contemplating the various ways in which the project might be unsuccessful (principle 4).

A certain tendency to exaggerate the merits of one's own approach and to play down contrary evidence may be necessary, particularly during the early stages of a project. Moreover, scientists like to be right and get recognition for being right. The satisfaction of demolishing a theory one has laboriously constructed may be small in comparison with the satisfaction of seeing it vindicated. All things considered, it's extremely hard for most people to adopt a consistently Popperian attitude toward their own work. In fact, part of the intellectual responsibility of a scientist is to provide the best possible case for important ideas, leaving it to others to publicize their defects and limitations. That is just what most scientists do (principle 5).

In a commencement address at Caltech some years ago, Richard Feynman endorsed the Popperian outlook by remarking, "It's a kind of scientific integrity, a principle of scientific thought that corresponds to a kind of utter honesty—a kind of leaning over backwards. For example, if you're doing an experiment, you should report everything that you think might make it invalid—not only what you think is right about it; other causes that could possibly explain your results; and things you've thought of that you've eliminated by some other experiment, and how they worked—to make sure the other fellow can tell

Figure 1.2
Richard Feynman,
Caltech commence-
ment, 1974. Courtesy
of Floyd Clark,
California Institute of
Technology Archives.

they've been eliminated."[4] That is a high-minded and laudable attitude to have, but it is far beyond the capacity of most scientists. Most scientists are content to present their results without calling attention to all the ways they could be wrong (principle 6). Nevertheless, it's important for scientists to be careful to point out what could be wrong if they know it.

It may be that merely verifying a hypothesis has little intrinsic value, but it is striking that the distribution of credit in science reflects a decidedly different view. Scientists win Nobel Prizes and other coveted accolades for detecting new effects or for predicting effects that are subsequently verified. It is only when a theory has become very well established that one receives significant credit for refuting it, and while such an achievement may burnish a scientist's reputation, it rarely, if ever, results in the type of rewards associated with an affirmative breakthrough. Unquestionably, rewarding confirmations over refutations provides scientists with incentives to confirm theories rather than to refute them, but as we have been arguing, that is not necessarily bad for science.

Conventional accounts of the scientific method share the assumption that all scientists should adopt the same strategies. In fact, government agencies used to define scientific misconduct as "practices that seriously deviate from those that are commonly accepted within the scientific community" (principle 7).[5] But rapid progress will be more likely if different investigators have different attitudes toward appropriate methods. As noted above, one important consequence of the winner-take-(nearly)-all-the-credit system is that it encourages a variety of perspectives, programs, and approaches. Thus, attempts to define misconduct in terms of deviations from commonly accepted practices are doubly misguided: Not only will such commonly accepted practices fail to exist in many cases, but also it will be undesirable to enforce the conformity that such a principle would require. More generally, we can see why attempts to discover "the" scientific method fail. There are deep, systematic reasons why all scientists should not follow some single, uniform method.

But that doesn't mean that "anything goes." The scientific community draws an important distinction between claims that are open to public assessment and those that are not, and a scientist who fabricates data will be judged far more harshly than one who merely extrapolates beyond the recorded data. The difference is that where there is no fabrication, nothing exists to obstruct the critical scrutiny of the work by peers. Since scientists must be able to trust that the data they are critiquing resulted from a legitimate experiment, fabrication of data is a far more serious violation of the scientific method than extrapolation.

Conducting an experiment in a way that produces reliable results is not just a matter of following rules. Experimenters, some more so than others, possess skills that allow them to get their experiments to work, often without even knowing what those skills are. Assessing whether a particular experimenter has produced reliable results may require a judgment based on whether she or he has produced dependable results in the past. The often essential but hard to quantify role of craftsmanship in designing and carrying out successful experiments is another reason why general rules of method have proved so elusive (principle 8).

These facts about specialization, skill, and authority have a number of consequences for understanding what constitutes proper scientific conduct. For example, behavior that strikes an outsider as exhibiting irrational deference to authority may have a serious rationale. When a scientist discards certain data on the basis of subtle clues in the behavior of the apparatus, and other scientists accept his or her judgment, this should not be attributed to the operation of power relationships (principle 9).

Another consequence has to do with the extent to which scientists are responsible for misconduct or sloppy research on the part of their collaborators (principle 10). It is precisely the point of many collaborations to bring together people from different specializations, with the implicit understanding that their different backgrounds and diverse abilities mean that they may not always be in the best position to accurately judge the quality of one another's work. Setting up a policy of holding scientists responsible for the misconduct of coauthors and coworkers would discourage a great deal of valuable collaboration.

Credit tends to go to those who are famous at the expense of those who are not. A paper signed by Nobody, Nobody, and Somebody will almost invariably be referred to as "work done in Somebody's lab." There are so many papers in so many journals that few scientists have time to read more than a fraction of those relevant to their work. Famous names tend to identify those works that are worth noticing. In certain fields, particularly biomedical fields, it has become customary to include the head of the lab as an author, even when the head of the lab didn't participate in the research. Some people refer to this practice as "guest authorship" and regard it as unethical (principle 11). However, the practice may be functionally useful and involve little deception, since it will be well known to all participants in a field. Physics is not such a field. Most physicists recoil at the thought of guest authorship.

This brings us to a view of science called the Ortega hypothesis. It is named after the Spanish philosopher José Ortega y Gasset, who wrote in his 1930 classic, *The Revolt of the Masses,* that "experimental science has progressed thanks in great part to the work of men astoundingly mediocre, and even less than mediocre. That is to say, modern science, the root and symbol of our actual civilization, finds a place for the intellectually commonplace man and allows him to work therein with success."

Ortega's assertion (principle 12) is probably based on the empirical observation that there are, in every field of science, many practitioners doing more or less routine work. Less empirically, it is also supported by the idea that knowledge of the universe is a kind of limitless wilderness to be conquered by the action of many hands relentlessly hacking away at the underbrush. An idea supported by both observation and theory has a very firm basis in science.

The Ortega hypothesis was named by two sociologists, Jonathan R. Cole of Columbia University and Stephen Cole of SUNY–Stony Brook, when they set out to demolish it in a 1972 article in *Science*. They wrote:

> It seems, rather, that a relatively small number of physicists produce work that becomes the basis for future discoveries in physics. We have found that even papers of relatively minor significance have used to a disproportionate degree the work of the eminent scientists.[6]

In other words, according to the authors, a small number of elite scientists are responsible for the vast majority of scientific progress. (The authors base these conclusions on their observations of the physics community, while contending that they are valid for all branches of science.) Seen in this light, the reward system in science is a mechanism that has evolved for promoting and rewarding the star performers.

If the Ortega hypothesis is correct, science is best served by producing as many scientists as possible, even if they are not of the highest quality (principle 13). However, if the elitist view is right, it is best to restrict production to fewer and better scientists. In any case the question involves ethical issues (What is best for the common good?) as well as policy issues (What is the best route to the desired goal?).

Scientific papers often misrepresent what actually happened in the course of the investigation(s) they describe. Misunderstandings, blind alleys, and mistakes of various sorts will fail to appear in the final written account. Nevertheless, the practice is nearly universal, because it is a more efficient means of transmitting results than an accurate historical account would be. Contrary to normal belief (principle 14), this type of misrepresentation is condoned and accepted in scientific publications, whereas other transgressions are harshly condemned. This practice may not be ideal, but it is an inherent way in which science is done.

Peer review has an almost mystical significance in the community of scientists. Published results are considered dependable because they've been peer reviewed, and unpublished data are not dependable because they have not been. (The last decade has seen a growing number of papers "published" in pre-press on the Web, without the advantage of peer review. These are naturally regarded as less reliable by most scientists.) Many consider peer review the ethical fulcrum of the whole scientific enterprise. For most small projects and nearly all journal articles, peer review is accomplished by sending the manuscript or proposal to referees whose identity will not be revealed to the authors.

The peer-review process is very good at separating real science from nonsense. Referees know the current thinking in a field and are aware of its rules and conventions. But it is not at all good at detecting fraud, as the cases of compromised papers that have successfully passed through peer review amply demonstrate (principle 15).

☞ It has become fashionable in recent decades for scholars from the social sciences and other disciplines to visit the exotic continent of Science and send back reports of their observations of the behavior and rituals of the natives. The resulting dialogues have not always been entirely amicable and have, in fact, sometimes been referred to as "the science wars." Let us extend an olive branch by offering an entirely unjaded, unbiased insider's view of this curious terrain.

There are undoubtedly many reasons why people choose to become scientists. Simple greed, however, is not high on the list. The reason is that the rewards for success in science are not primarily monetary (although a certain degree of material well-being does often follow in their wake). If you are a scientist, each success is rewarded by the intoxicating glow that comes from knowing or believing that you have won at least one small round in the endless quest for knowledge. That glow fades quickly, however, unless it brings with it the admiration and esteem of your peers and colleagues (who are, after all, the ones capable of understanding most fully what you have done and are frequently the only ones who care). The various means by which scientists express their admiration and esteem for their colleagues are so subtle and complex that they beggar the etiquette of a medieval royal court. We will call these means collectively the Reward System of science.

Closely linked to the Reward System is a second organization that we may call the Authority Structure. The Authority Structure guides and controls the Reward System. Moreover, certain positions within the Authority Structure are among the most coveted fruits of the Reward System. Nevertheless the two are not identical. The pinnacle of the Reward System is scientific glory, fame, and immortality. The goal of those in the Authority Structure is power and influence. Scientists distinguish sharply between the two. They will sit around the faculty lounge or the lunch table lamenting the fate of a distinguished colleague who has become the president of a famous university. "He was still capable of good work," they will say, sounding much like saddened warriors grieving the fate of a fallen comrade. The university president is a kingpin of the Authority Structure but a dropout from the scientific Reward System.

The Reward System and the Authority Structure are both rooted in the institutions of science. These institutions vary somewhat from one discipline to another and from one country to another, but the broad outlines will be recognizable to all. Our discussion is most influenced by the physical sciences as they are practiced in the United States, but it will apply broadly to all science, in all countries.

Scientific research is performed in universities, and to a lesser extent in colleges that do not grant doctoral degrees. It is also performed in national laboratories and in industrial laboratories. The universities and colleges may be public or private. The national laboratories may be run directly by government agencies or managed for the government by universities or consortia of universities. Industrial laboratories are usually, but not always, operated by a single company.

Scientific societies, such as the American Physical Society or the American Chemical Society, have members from all of the above types of scientific institutions. The societies organize national and regional scientific meetings, publish journals, and administer the awarding of certain prizes and honors. They are private organizations, whose officers are elected by their members and whose costs are paid by the dues of their members and by other related sources of income. There are a few scientific societies (such as the American Association for the Advancement of Science) that are not tied to a particular scientific discipline but still hold meetings and publish journals.

There are also purely honorary societies, typified by the National Academy of Sciences (NAS). The NAS holds meetings, publishes a journal, and serves certain needs of the government through its research and consulting arm, the National Research Council. However, by far the most important thing the NAS does is to elect its own members. Election to the NAS is one of the highest rungs on the Reward System ladder.

These are the elements of the institutions of science. We have left out a few crucial items, such as the Scandinavian bureaucracy (the Royal Swedish Academy of Sciences and the Royal Caroline Institute) that awards Nobel Prizes, and the inscrutable college of historians and journalists that somehow decides which scientists shall become famous outside of science itself. However, even within the elements described, there are infinitely subtle layers of influence and prestige.

Behind a carefully cultivated veneer of cordiality, colleges and universities wage a fierce, endless struggle of titanic proportions for positions of honor in a peculiar contest. No one is quite sure who's keeping score, but everyone knows roughly what the score is. The contest ranks each university against others, each college against others, and within a single discipline, departments against one another. (Similar rivalries exist among national laboratories, industrial laboratories, and even federal funding agencies.)

To the aspiring academic scientist, the steps on the perilous ladder to fame and glory look something like this:

1. Be admitted to a prestigious undergraduate college or university (useful but not essential).
2. Graduate with a B.S. degree (essential).
3. Be admitted to a prestigious graduate department (very important).
4. Graduate with a Ph.D. (essential).
5. Get a postdoctoral appointment or fellowship at another prestigious university (this almost always ranks lower in

the invisible hierarchy than the university where you did your graduate work).

6. Get a position as assistant professor. The caliber of the university and department is crucial, since you are unlikely ever to move up from there in the invisible rankings. National and industrial laboratories also have positions analogous to assistant professor, and some people prefer the risky course of starting in an industrial lab with the hope of being successful enough to be called to a university later.

7. Bring in outside research support (mostly from federal agencies), attract graduate students of your own, get papers published in the best journals (that usually means the ones published by the professional societies—but there are exceptions, such as *Nature*, which is privately published), get invited to speak at national or (even better) international meetings sponsored by professional societies, and generally become visible among active scientists in your field outside your own institution. It is useful, but not essential at this stage of your career, to teach well and to participate in academic committees and the like. All of these demanding and challenging steps are to be taken honestly, without the remotest hint of scientific misconduct or fraud.

8. Get tenure (as a result of doing number 7 very well).

9. Get promoted to full professor.

10. Your colleagues darkly suspect that you will now rest on your laurels, and you must prove them wrong. Get more funding; expand the size of your research group (graduate students, postdocs, technicians, etc.). Get yourself

appointed to national boards, panels, and committees, secure more invitations to speak at more meetings, and so on. If at all possible, get something (a discovery, a technique, a program, and a piece of hardware are all acceptable options) named after yourself. This is the most effective way of getting noticed, but it's also tricky, since someone else must do it for you, and then it has to catch on among workers in the field. Once again, there must be not the slightest whiff of scientific misconduct here. You might do all of these things motivated purely by the thrill of discovery, but do them you must.

11. The following are now available if you work hard enough to get them and manage to have a little luck in your research:

Awards and prizes from your professional society,

A named professorship,

Membership in a National Academy,

Major national and international prizes up to the Nobel itself, and

Immortality.

At each of these various steps, you have faced gatekeepers from the Authority Structure of science. They are generally people who have ascended a few rungs above that level but then stepped out of the competition (remember the university president mentioned earlier). For example, the faculty of an undergraduate college (where you may choose to attempt steps 1 and 2) will generally have reached step 4 (a Ph.D.), and perhaps 5 (a postdoc), but opted out of the research competition at step 6 (by taking a position in a college rather than a research university). They may very well never have intended to climb any higher than necessary to reach their positions as college faculty, but it would have been unwise for them to admit as much while they were climbing. Each of the gatekeepers they faced probably had to be convinced that they were aspiring to the very pinnacle. These are the people who will now decide your fate. They are most likely to be impressed if they believe you are aspiring to that same pinnacle.

At the graduate school level, your Ph.D. thesis advisor, a very important person in your life, will probably (had better be) still climbing and may very well have climbed quite high already, but decisions about you will be made also by department chairs, deans, and others who have traded their places on the ladder for positions in the Authority Structure of science.

Once you pass the Ph.D. hurdle, the rules for scaling successive steps become increasingly less well defined. The rules are often unwritten, and the people you must impress are further afield. Each promotion will require confidential letters of recommendation from people outside your own institution, solicited not by you but by the chair of a committee. You will thus be expected to be known by people you have not met, merely because of your growing scientific reputation. Your reputation will be based on published papers whose fate will be in the hands of journal editors and anonymous referees chosen by them. The research reported in those papers will be possible only if you can win financial support on the basis of research proposals submitted to the granting agencies. Your proposals will be handled by project officers (either permanent or temporary refugees from the race up the research ladder) and judged once again by anonymous referees or a panel of active scientists. Finally, even if you manage to finance and publish your work, it will be little noticed unless you manage to get invited to speak at national meetings organized by your professional society. The staff of the society will generally have dropped out of the race, but decisions about who speaks will most likely be made by committees of active scientists.

Notice that at each point of decision, there tend to be two kinds of gatekeepers. One kind is an administrator (department chair or dean, journal editor, project officer, professional society staff) and the other kind an active scientist (writer of letters of recommendation, anonymous referee, member of panels and committees). The first kind of gatekeeper has often stepped out of the race (the position itself is generally the reward for having reached a certain level), while the other is still very much in the race. The people in this latter group are not only your judges, they are also your competition. Furthermore, you have become one of them. People in the other group, if they are no longer in competition with you, have often forgotten the fierce struggle you face, and moreover they tend to have the curious view that you are working for them.

It should be clear from this discussion that scientific score-keeping is no simple matter. The issue of who will emerge as successful and famous in science depends in large measure on who has the best ideas and who works the hardest. In that sense, science is a true meritocracy. However, there are very clearly other elements at play here. One of the most important is being in the right place at the right time. For example, the discovery of quantum mechanics early in the twentieth century swept a whole generation of theoretical physicists to fame and glory. The very best made truly fundamental contributions, but even those of more modest talent found untouched problems ready to be solved with the new theory. Another example is supplied by World War II's mega–science projects, chief among them the Manhattan Project and MIT's Radiation Lab, which swept yet another generation of physicists to power and influence.

In addition to the factors we have just outlined, there are others that have been observed and documented, that arise out of the behavior and customs of scientists as a group. The late sociologist Robert K. Merton called one of them the Matthew effect, following this passage in the Gospel according to Matthew: "For unto every one that hath shall be given, and he shall have abundance: but from him that hath not shall be taken away even that which he hath."[7]

The Matthew effect in science is the observation that credit tends to go to those who are already famous, at the expense of those who are not. For example, if a paper is written by a team of researchers, only one of whom is well known in the field, readers will tend to refer to the article by the alpha scientist's name even if it is far back in the authorial pack.

The roots of this scientific Reward System and the Authority Structure date back to the seventeenth century, almost to the birth of modern science itself. It is probably fair to say that experimental physics was invented by Galileo Galilei (1564–1642), who discovered the law of falling bodies and the law of inertia by means of experiments using ingeniously crafted instruments. The scientific research laboratory was first created not much later, by English chemist Robert Boyle, who set up a team of assistants, specialists, technicians, and apprentices to carry out systematic chemical investigations. Both Galileo and Boyle belonged to scientific societies that still exist (*L'Accademia dei Lincei* and the Royal Society, respectively). Boyle supported his research by means of his own wealth, but Galileo spent much of his time and energy seeking what we would today call government and private sponsorship. (It is not for nothing that the discoverer of the moons of Jupiter named them the *Sidera Medicea*—the "Medicean stars." Patronage by the Medici no

Figure 1.3
Galileo Galilei. Photo reproduction of Robert Hart's stipple engraving published by Charles Knight of London in 1834, after a 1757 oil on canvas portrait done by Scottish portrait painter Allan Ramsay (1713–84) and presented to Trinity College, Cambridge, in 1759, where it hangs in the Master's Lodge; Ramsay was inspired by Flemish portraitist Justus Sustermans's (1597–1681) oil on canvas portrait of Galileo painted circa 1640, which hangs in the Pitti Gallery in Florence. Courtesy of California Institute of Technology Archives.

longer being what it once was, they are today more commonly called the Galilean satellites.) Both Galileo and Boyle also engaged in fierce struggles with others over priority for scientific discoveries. In other words, the basic outlines of the social organization of science emerged almost as soon as science did, and it was firmly in place by the time Isaac Newton (who became a named professor at Cambridge and the first president of Great Britain's Royal Society) wrote his *Principia*. It is difficult to avoid the conclusion that science cannot exist—and certainly cannot flourish—without the Reward System and the Authority Structure.

Of course, professional societies, prizes, and awards, to say nothing of department chairs and deans, are by no means limited to the sciences. One can detect the basic elements of the Reward System and the Authority Structure in virtually every academic discipline. Nevertheless, it seems better developed and more highly organized in the sciences than elsewhere. The reason is undoubtedly to be found both in the nature of science and in human nature, since it is we humans who must pursue science. Science is basically a collaborative enterprise to discover important truths about the world, carried out by individuals who are generally more strongly motivated by their own interests than by the collective good. The Reward System and the Authority Structure serve to regulate and channel this collaboration-*cum*-competition to produce useful results. So long as it succeeds in doing so, this system of ours seems likely to remain firmly in place.

In all of this, scientific misconduct plays a peripheral role, lurking quietly in the shadows: a temptation, perhaps, for some at each stage, but never a central point. The mountains described here must be scaled without a hint that any untoward activity has contributed to the ascent. Anything else is utterly unacceptable—but, as we are about to discover, not always unthinkable. With that in mind, we turn to some illuminating episodes in the history of modern science.

Two
In the Matter of Robert Andrews Millikan

Robert A. Millikan was a founder, first leader, first Nobel laureate, and all-around patron saint of the California Institute of Technology (Caltech). In his day, Millikan was a lionized titan of American science, but his reputation has received critical scrutiny in recent times.[1] He has been accused of male chauvinism, anti-Semitism, mistreating his graduate students, and—last but not least—scientific fraud. We shall now look into these charges.

Millikan was born in 1868, the son of a Midwestern minister. He attended Oberlin College, got his Ph.D. in physics from Columbia University, did some postdoctoral work in Germany, and in the last decade of the nineteenth century took a position at the brand-new University of Chicago in a physics department headed by his idol, A. A. Michelson.

During the next decade, Millikan wrote some very successful textbooks, but he made little progress as a research scientist. This was a period of crucial advancement in physics. J. J. Thomson discovered the electron; Max Planck kicked off the quantum revolution; Albert Einstein produced his theories of special relativity and the photoelectric effect; and Einstein's theoretical work and French physicist Jean Perrin's experiments on Brownian motion established conclusively that matter was made of atoms. Professor Millikan made no contribution to these events. Nearing forty years of age, he became very anxious indeed to make his mark in the world of physics. He chose to try to measure the charge of the electron.

The electron, the fundamental unit of electricity, had been speculated about for decades, but no one had succeeded in proving its existence. Cathode-ray tubes had been around for even longer when, in 1896, the great British experimentalist J. J. Thomson succeeded in showing that all cathode rays are electrically charged and have the same ratio of electric charge to mass.[2] Such a ratio could exist only if the charge was carried by discrete particles; as a result, Thomson's finding was widely regarded at the time as constituting the discovery of the electron as well as providing the first demonstration that atoms had internal parts. (That the atom had a nucleus and that the nucleus contained protons and neutrons would have to wait until the discoveries of Ernest Rutherford and others later on.) Although Thomson called his new, negatively charged particle a "corpuscle," the name "electron" had already been coined, and it was the one that became permanent. The challenge then was to measure the new particle's electric charge independent of the mass. At the University of Cambridge, Thomson and his colleagues set out to do that in a series of cloud chamber experiments.

Although they have been largely superseded by far more sophisticated (and considerably more expensive) particle detectors, cloud chambers—devices in which a sudden decompression makes tiny water droplets form on electrically charged ions—were once a mainstay of experimental physics. By applying an electric field and observing how it affected the rate at which the ionized clouds of water droplets drifted downward inside the chamber, Thomson hoped to find the charge on a single electron. He singled out the upper edges of the clouds as the likeliest place to look, because the droplets that formed there were generally too small to hold more than one charge. In this way, a crude but correct estimate of the charge on the electron could be obtained.

These cloud chamber experiments were the starting point of Millikan's efforts. Working with one of his University of Chicago graduate students, Louis Begeman, he had the idea of applying a much stronger electric field than any used previously. His reasoning was that this would completely halt the falling cloud, allowing him to make a much more precise measurement of the charge on its droplets. To his surprise, when he trained a laboratory telescope (not to be confused with an astronomical instrument) on the cloud in order to observe the drops, he found that applying a powerful field had not stopped the cloud's descent. Instead, nearly all of the ionized droplets dispersed upward or downward, leaving in view just those few droplets whose electric force came sufficiently close to balancing the effect of gravity to temporarily immobilize them. Millikan quickly realized that measuring the charge on individual ionized droplets was a method far superior to finding the charges on droplets in a cloud.

It may have been during this period that Millikan's wife, Greta, attending a social event while Millikan spent one of his many long evenings in the lab, was asked where Robert was. "Oh," she answered, "He's probably gone to watch an ion." "Well," one of the faculty wives was later overheard to say, "I know we don't pay our assistant professors very much, but I didn't think they had to wash and iron!"[3]

Unfortunately, with water as the medium, the single-droplet method had a serious flaw: The water evaporated too rapidly to allow accurate measurements. Millikan, Begeman, and a new graduate student named Harvey Fletcher decided to try doing the experiment with some substance that evaporated less rapidly, and Millikan assigned to Fletcher the job of devising a way to use mercury or glycerin or oil. Fletcher chose oil and almost immediately got a crude apparatus working, using tiny droplets of watch oil that could be sprayed out of a perfume atomizer he had bought in a drugstore. When he focused his telescope on the suspended oil droplets, he could see them dancing around as they collided with unseen air molecules, in a textbook demonstration of Brownian motion. This itself was a phenomenon of considerable scientific interest.

When Fletcher finally persuaded the busy Millikan to look through his telescope at the dancing, suspended droplets of oil, Millikan immediately abandoned all his efforts to pull off the experiment with water and turned his attention to refining the oil-drop method.

It took a couple of years, but by around 1910 Fletcher and Millikan had produced two results. The first was an accurate measurement $(4.774 \times 10^{-10}$ electrostatic units) of the unit electric charge (called e), which they determined from observing how rapidly or how slowly the oil drops rose or fell in gravitational and electric fields. The second result was the determination of Ne, a crucial variable related to Brownian motion.

Millikan then approached his student Fletcher with a deal, predicated on the notion that at least two significant articles were likely to emerge from these experimental results. Fletcher could write one of them and use it as the basis for his Ph.D. thesis, provided that he was sole author on the one paper while Millikan enjoyed sole authorship of the other. In this inspired division of labor, Millikan proposed that Fletcher write up the Brownian-motion work while he, Millikan, took custody of the unit electric-charge work. This is the source of the assertion that Millikan mistreated his graduate students. No doubt

Millikan understood that the measurement of e would establish his reputation, and he wanted the credit for himself. Fletcher understood this as well as Millikan. He was somewhat disappointed, but Millikan, his protector and champion throughout his graduate career, had in effect made him an offer he could not refuse. The two men remained good friends throughout their lives, and Fletcher, who went on to become a distinguished researcher at Bell Labs, saw to it that this version of the story was not published until after both he and Millikan were dead.[4]

Let us turn now to the question of scientific fraud. In 1984, the research honor society Sigma Xi published a booklet called *Honor in Science*. More than a quarter of a million copies were distributed before it was replaced by a revised version. The original edition includes a brief discussion of the Millikan case that begins, "One of the best-known cases of cooking is that of physicist Robert A. Millikan."[5] Cooking, which in research nomenclature means retaining only those results that fit the theory and discarding others, is one of the classic forms of scientific misconduct, first described in a book by computer pioneer Charles Babbage in 1830.[6] According to *Honor in Science*, reliable and well-informed authorities (of whom more later) had found Millikan guilty of cooking the data that ultimately led to his Nobel Prize.

Do these charges have any validity? There are really two issues. One is the question of what actually happened between 1910 and 1917. The other concerns how Millikan, much more recently, came to be accused, tried, and convicted of scientific misconduct. Let us now consider each of these in turn.

The accusation of fraud against Millikan, very briefly, consists of this: Millikan's article presenting his measurement of the unit of electric charge was published in 1910, and shortly thereafter Millikan found himself embroiled in a controversy with the Viennese physicist Felix Ehrenhaft. Using an apparatus rather similar to Millikan's, Ehrenhaft claimed to have found cases of electric charges much smaller than Millikan's value of e and to have further determined that the overall electrical charge invariably fluctuated within a range of disparate values. If true, this assertion would have effectively demolished Millikan's claim to have accurately measured the *fundamental* unit charge of the electron, with the further implication that as electricity was not composed of discrete unit charges, the electron did not in fact exist.

Although he dismissed Ehrenhaft's observations of smaller charges as erroneous, Millikan nevertheless found it prudent to name them, referring to them in a subsequent publication as "subelectrons." To refute Ehrenhaft's assertion, Millikan (now working alone; Fletcher had gotten his doctorate and left) made a new series of measurements, in which he set out to demonstrate that the charge on every single oil droplet he studied equaled the value he had experimentally established for e or was, within a very narrow range of error, an integer multiple of a single value of e. In other words, having arrived at a precise value for a droplet containing a single charge (e), Millikan now sought to demonstrate that larger charges were always two, three, or four times (and so forth) the value of e and therefore came from droplets that contained two, three, or four (or more) units of charge. In 1913, he published his findings in a paper whose clarity and stylistic elegance succeeded in dispatching Ehrenhaft (who, in any case, had no particular standing within the scientific community) and which contributed significantly to Millikan's being awarded the Nobel Prize in Physics.[7]

However, an examination of Millikan's private laboratory notebooks (housed in the Caltech Archives) reveals that he did not in fact report the data he recorded from every oil droplet.[8]

He reports the results of measurements on 58 drops, whereas the notebooks show that he took data on approximately 175 drops between November 20, 1911, and April 16, 1912. In a classic case of cooking, the accusation goes, he reported results that supported his own hypotheses and discarded those results that would have supported Ehrenhaft's position. And, to make matters very much worse, he lied about it. His 1913 paper contains this explicit assertion: "*It is to be remarked, too, that this is not a selected group of drops but represents all of the drops experimented upon during 60 consecutive days* [italics in the original], during which time the apparatus was taken down several times and set up anew." Thus, Millikan stands accused of cheating and of covering up his cheating by lying about it in one of the most important scientific papers of the twentieth century. Could there be a clearer case of scientific misconduct?

To arrive at an answer, let us look at some of the pages in Millikan's private laboratory notebooks. Figure 2.1 shows a page dated November 18, 1911.

At the top right is written the temperature, $t = 18.0°C$, or about 66°F. (Obviously, Millikan's lab was not well heated against the bitter Chicago weather; and the barometric pressure, 73.45 cm, suggests it may have been a stormy day—76 cm is an average reading.) On the left, a column of figures has been listed under G, for gravity. These represent the time it took each tiny droplet—pinpoints of light too small to focus in Millikan's laboratory telescope—to fall between scratch marks in his telescope's focal plane. These figures give the terminal velocity of the drop when the viscosity of air balances the force of gravity. From this measurement alone, Millikan could determine the size of the tiny, spherical drop. Then there is a second column headed F, for field. The figures in this column denote the amount of time

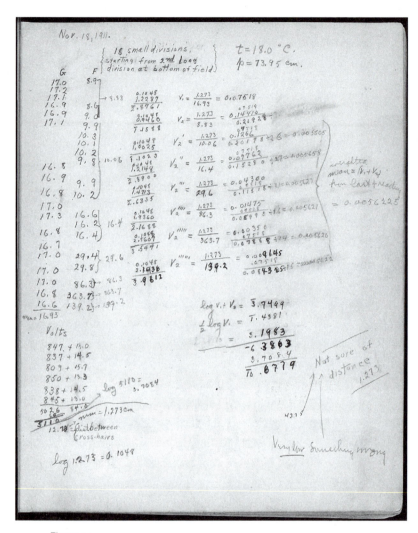

Figure 2.1
Millikan's laboratory notebook page, Chicago, Saturday, November 18, 1911. Courtesy of California Institute of Technology Archives, Robert Andrews Millikan Collection.

it took each drop to rise between the scratch marks under the combined influence of gravity, viscosity, and the applied electric field, which had been turned off during the "G" measurements. In somewhat simplified terms, subtracting the G (gravity + velocity) from F (gravity, velocity, and charge) gives the value for charge alone.

Continuing to follow Millikan's entries, we can see that the value for the F measurements changes from time to time. The first series gives an average of 8.83, then 10.06, then 16.4, and so on. That happens because the charge on the drop changes every so often, when the drop captures an ion from the air. If a negatively charged drop captured a positive one, it might have turned neutral; if it captured the same charge (+ or −), it would acquire a larger charge, providing Millikan with the data he needed to show that larger charges were always an integer multiple of e. To the right of these columns on the same page, we find a series of laborious hand calculations (not necessarily done on the same day as the data were taken), using logarithms to do multiplication and square roots.

Then finally, at bottom right, comes this comment: "[charge] very low something wrong," with arrows leading to "not sure of distance. . . ." Needless to say, this apparently sub-charged drop was not one of the fifty-eight drops Millikan published.

The next figure (fig. 2.2) shows observations of two drops on November 20 and 22, with similar columns of figures. To the right at the bottom of the first observation, we see again "very low something wrong" and below that, "found meas[uremen]t of distance to the hole did not. . . ." Once again, the value of e was not up to snuff. But on the third figure (fig. 2.3), a page dated "Wednes. Dec. 20, 1911" (the temperature is now recorded at a comfortable 22.2°C—did the university turn the heat on in

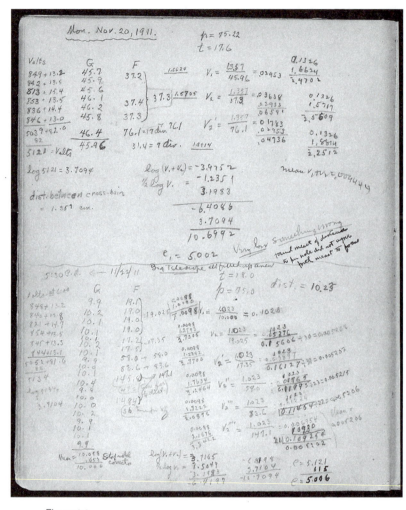

Figure 2.2
Notebook page, Monday, November 20, 1911. Courtesy of California
Institute of Technology Archives, Robert Andrews Millikan Collection.

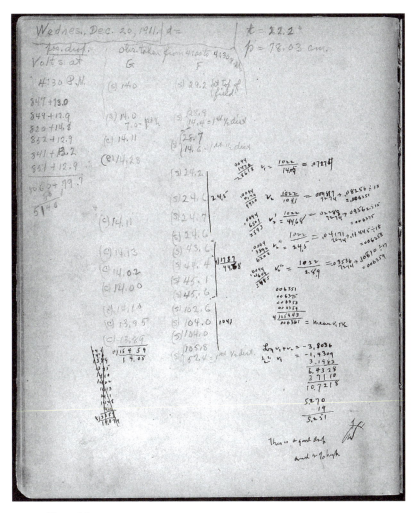

Figure 2.3
Notebook page, Wednesday, December 10, 1911. Courtesy of California
Institute of Technology Archives, Robert Andrews Millikan Collection.

December?), we find the remark, "This is almost exactly right, the best one I ever had!!!"

It is important to note that in his crucial 1913 paper, Millikan did not publish any of the drops for which the raw data are shown in these first three figures, not even "the best one I ever had." The omission of even this impeccable drop makes it clear that these initial experiments were all part of a warm-up period. During that time, Millikan gradually refined his apparatus and technique in order to make the best measurements anyone had ever made of the unit of electric charge.

The first observation that passed muster and made it into print was taken on February 13, 1912, and all of the published data were taken between that date and April 16, a period of sixty-three days (1912 was a leap year). Raw data taken during this period are shown in the fourth figure, dated March 14 (fig. 2.4). Our eye is immediately drawn to the comment on the top center part of the page, "Beauty Publish." Note also the barometric pressure, which is 16.75 cm, too low for even the stormiest day in Chicago.

Between February 13 and April 16, Millikan recorded data for about a hundred separate drops in his notebooks. Of these observations, about twenty-five were obviously aborted during the run and so cannot be counted as complete data sets. Of the remaining seventy-five or so, he chose fifty-eight for publication. Millikan's standards for acceptability were exacting. If a drop was too small, he did not carry out a calculation of its charge, on the assumption that it would be excessively affected by Brownian motion, or at least by inaccuracy in Stokes' law for the viscous force of air (more about this later). If the drop was too large, he also rejected it, reasoning that it would fall too rapidly for accurate measurement. As noted earlier, he also preferred to

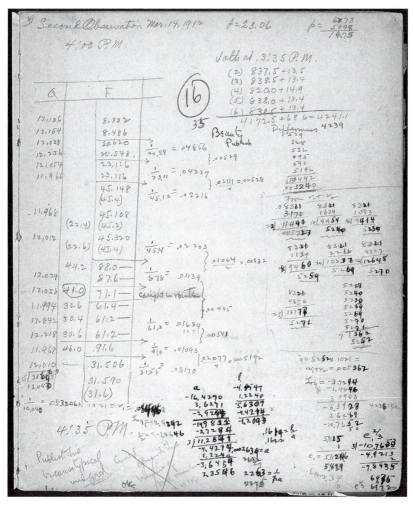

Figure 2.4
Notebook page, second observation, March 14, 1912. Courtesy of California Institute of Technology Archives, Robert Andrews Millikan Collection.

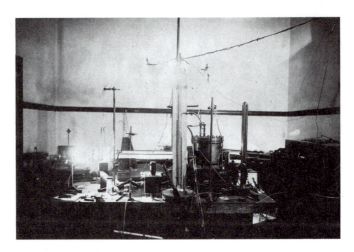

Figure 2.5
Millikan's oil-drop apparatus, Chicago, ca. 1912. Courtesy of California
Institute of Technology Archives.

have a drop change its charge a number of times in the course
of an observation. With these stringent criteria, it should not be
surprising that Millikan chose to use the data on only fifty-eight
of the drops that he observed during the period when he and his
apparatus had reached near perfection.

Furthermore, he had no special bias in choosing which
drops to discard. A modern reanalysis of Millikan's raw data by
historian of science Allan Franklin reveals that Millikan's result
for the unit of charge and for the limits of uncertainty in the
result would barely have changed at all had he made use of all
the data he had, rather than just the fifty-eight drops he used.[9]

I don't think that any scientist, having studied Millikan's
techniques and procedures for conducting this most demanding
and difficult experiment, would fault him in any way (and in fact
none has) for picking out what he considered to be his most

dependable measurements in order to arrive at the most accurate possible results. In the 1913 paper, he cites his result with an uncertainty of 0.2 percent, some fifteen times better than the best previous measurement (which reported an error of 3 percent). Furthermore, the most recent measurements of the charge on the electron, using the most accurate instrumentation available, are within Millikan's stated uncertainty of 0.2 percent.

In sum, Millikan's experiment was nothing less than a masterpiece, and the 1913 paper reporting it is a classic of scientific exposition. But there is still the matter of the damning statement that "this is not a selected group of drops but represents all of the drops experimented upon during 60 consecutive days," which is manifestly untrue.

The question is, why did Millikan mar his rigorous and elegant experiment with what was apparently an outright lie? In 1978, many years after the fact, Millikan's work was studied by historian of science Gerald Holton, who told the story of the Millikan-Ehrenhaft dispute and contrasted Millikan's published results with what he found in Millikan's laboratory notebooks.[10] Holton did not accuse Millikan of misconduct of any kind but instead found in the unpublished notebooks an opportunity to contrast a scientist's public, published behavior with what went on in the privacy of his laboratory.

Holton's work was seized upon by two journalists, William Broad and Nicholas Wade, who in 1982 published a book about misconduct in science called *Betrayers of the Truth*. Broad and Wade, both of whom were then reporters for *Science* magazine and now write for the *New York Times*, emerged as the chief proponents of the notion that Millikan was guilty of scientific misconduct, and their contention formed the basis for subsequent accusations, including the one published in Sigma Xi's *Honor in Science*.

In *Betrayers of the Truth,* Broad and Wade set out to make the point that scientists cheat. Their second chapter, "Deceit in History," commences with a roll call of culprits: Claudius Ptolemy, Galileo Galilei, Isaac Newton, John Dalton, Gregor Mendel, and Robert Millikan. At the very least, Millikan is in good company. Of Millikan they say he "extensively misrepresented his work in order to make his experimental results seem more convincing than was in fact the case." In a classic case of cooking, their accusation goes, he reported results that supported his own hypothesis regarding an exact value of the charge of the electron and ignored any data that seemed to bear out Ehrenhaft's observations regarding subelectrons.

These assertions are profoundly incorrect. (Incidentally, the accusations against most of the other scientists on the above list are equally spurious.) For Broad and Wade's arguments to make sense, the whole point of Millikan's exercise would have to have been to prove that there exists a smallest unit of electric charge and that subelectrons did not exist, because nothing else could have caused him to select data in a biased fashion. We would have to imagine that the existence of electrons (and by implication the existence of atoms) was an issue of burning controversy in 1913, with Millikan on one side of the great divide and Ehrenhaft on the other. In fact, there still were in 1913 a small number of respectable scientists who persisted in regarding the existence of unseen atoms (and electrons) as an unnecessary and unscientific hypothesis, but by then they had drifted far from the scientific mainstream—and even they would not have chosen Ehrenhaft as their champion, although his claims concerning subelectrons certainly suited their views. To Millikan, who had observed Brownian motion with his own eyes, the existence of atoms and electrons was beyond dispute.

Every revision of his technique, every improvement of his apparatus, every word he wrote, public or private, was directed to one goal only: the most accurate possible measurement of the fundamental charge of the electron. Ehrenhaft and the supposed controversy are never so much as mentioned in his landmark 1913 publication.

It is also worth remembering that history has vindicated Millikan, in that his result is still regarded as correct. Nevertheless, we are still stuck with the compromising *published* assertion that his data represent "all of the drops experimented upon during 60 consecutive days." To understand that statement in its proper context, we must make a small digression.

Millikan's oil drops rose and fell under the influence of three countervailing forces: gravity, electricity, and viscosity. The behavior of the first two under controlled laboratory conditions was very well understood. For the third force, viscosity, the nineteenth-century hydrodynamicist George Stokes had produced an exact formula applicable to a sphere moving slowly through an infinite, continuous viscous medium. The conditions that would conform exactly to Stokes' law were well satisfied by Millikan's oil drops in all respects except one: The drops were so small that the air through which they moved could not, from the minuscule drops' perspective, safely be considered a continuous medium. As we know today (and as Millikan firmly believed), the air was made up of molecules, and, compared to the size of an oil drop, the average distance between these molecules was not negligible. For this reason, Stokes' law could not be depended on to provide a complete description of Millikan's observations regarding viscosity.

To deal with this problem Millikan assumed—entirely without theoretical basis, as he stressed in his published paper—that the deviation from Stokes' law could be adequately corrected by the introduction of an additional unknown term that was strictly proportional to the ratio of the distance between air molecules to the size of the drop, so long as that ratio was reasonably small. To test this idea he purposely made his theoretical ratio larger than it had to be, by pumping some of the air out of his experimental chamber, which by decreasing the number of air molecules increased the distances between them. That is the reason he recorded such low pressure in that March 14, 1912, page from his notebook.

Then, when he had assembled all of his data, he used a trick that would be appreciated by any experimentalist. He plotted a graph of all his data for the fifty-eight drops on which he reported. This was done in such a way that if his supposition regarding Stokes' law was correct, all the data points would fall on a single straight line, and the position of the line on the graph would give the value of the unknown correction term. A successful plot would simultaneously prove that his method of correcting Stokes' law was justified and provide the magnitude of the necessary correction. As with everything else in this experiment, this approach was not intended to investigate whether an electron's charge came in units (Millikan accepted this without question) but rather to measure the unit of charge with the greatest possible accuracy.

Now let us turn to Millikan's actual published paper, which begins on page 109 of volume II, series 2 of *The Physical Review*. He first provides a detailed summary of how he has conducted his experiment, and then on page 133 he writes: "Table XX. contains a complete summary of the results obtained on all of the 58 different drops upon which complete series of observations like the above were made during a period of 60 consecutive days."

As we have already seen, his published results came from measurements made over a period of sixty-three, not sixty, days, but I think we can forgive him that lapse. The clear implication of the sentence is that there were only fifty-eight drops for which the data were complete enough to be included in the analysis. Page 133 is followed by two pages of Table XX. and by an additional two pages of the graph testing the correction to Stokes' law. On page 138, Millikan points out that all of the points do indeed fall on the line and that in fact "there is but one drop in the 58 whose departure from the line amounts to as much as 0.5 percent."

The now notorious sentence follows: "It is to be remarked, too, that this is not a selected group of drops but represents all of the drops experimented upon during 60 consecutive days. . . ." However, the incriminating remark is made not in regard to whether charge comes in units but in regard to getting the correction to Stokes' law right. In effect, he is saying, "Every one of those 58 drops I told you about confirms my presumed formula for correcting Stokes' law." Although this sentence makes its appearance five pages after the fifty-eight-drop remark on page 133, the intervening pages are tables and graphs. It seems most likely that in Millikan's original typescript (which does not survive, as far as we know) that sentence would have almost immediately followed the reference to the number of drops. In other words, a careful reading of Millikan's words in context greatly diminishes their apparent significance as evidence of misconduct.

In 1917, when Millikan published his book *The Electron*, he did take the trouble to confront Ehrenhaft explicitly and very effectively demolish Ehrenhaft's arguments.[11] He also used verbatim the section of his 1913 paper on Stokes' law, thus repeating the offending assertion that he had reported data for

Figure 2.6
Robert A. Millikan
and his wife, Greta,
1953. Courtesy of
California Institute
of Technology
Archives.

every drop he tested. Most probably, by 1917 he had forgotten the very existence of the other drops he had observed, however incompletely, between February and April 1912.

Millikan's real rival was never the hapless Ehrenhaft but J. J. Thomson. This was not because they disagreed scientifically but because both wanted to be remembered in history as the father of the electron. As things turned out, there was more than enough credit to go around.

In recent years, Millikan's life and pronouncements have become a juicy target for certain historians, partly because he was white, male, very much a part of the establishment, and, of course, no longer here to defend himself. On the issue of sexism, for example, there is a letter in which Millikan advised the president of Duke University, W. P. Few, not to hire a female professor of physics.[12] By the time he wrote this letter, in 1936, Millikan was the renowned and powerful leader of the California Institute of Technology (he never accepted the title of president, and was known throughout his tenure as chairman of the executive council). Few had written to Millikan in confidence, asking his advice on this delicate issue.

Millikan's reply shows his unease: "I scarcely know how to reply to your letter," he begins. Acknowledging in a later paragraph that "women have done altogether outstanding work and are now in the front rank of scientists in the fields of biology and somewhat in the fields of chemistry and even astronomy," he then cautions Few that "we have developed in this country as yet no outstanding women physicists." He goes on to say that "Fraulein Meitner in Berlin and Madame Curie in Paris" are among the world's best physicists, but points out that they live in Europe, not the United States. "I should therefore," he concludes his confidential advice, "expect to go farther in influence and get more for my expenditure if in introducing young blood into a department of physics I picked one or two of the most outstanding younger men, rather than if I filled one of my openings with a woman."

In his private correspondence, Millikan also reveals a stereotyping of Jews. Writing from Europe to his wife, Greta, in 1921, he describes the renowned physicist Paul Ehrenfest as "a Polish or Hungarian Jew [Ehrenfest was in fact Austrian] with a very short stocky figure, broad shoulders and absolutely no neck. His suavity and ingratiating manner are a bit Hebraic, (unfortunately) and to be fair perhaps I ought to say too that his genial open-mindedness, extraordinarily quick perception and air of universal interest and inquiry are also characteristic of his race."[13]

What are we to make of these comments? They are certainly not the rantings of a mindless bigot. Undoubtedly Millikan's biases were typical at the time of a man of his upbringing and background. It should also be said that whatever prejudices he harbored, they never interfered with his judgment of scientists. His hero A. A. Michelson was Jewish, as were many of the stellar faculty Millikan personally recruited to Caltech: theoretical physicist Paul Epstein, Albert Einstein, fluid dynamics and aerospace pioneer Theodore von Kármán, and the outstanding seismologist Beno Gutenberg, among others. On the distaff side, the evidence is perhaps more mixed: Caltech was an all-male school in Millikan's day and remained so for many years after his death.

📐 Three
Bad News in Biology

In the late 1980s, Caltech, one of the nation's premier research institutions, was stunned when two apparently unrelated instances of research misconduct showed up in the laboratory of one of its star performers, biologist Leroy E. Hood. Accused of misconduct were Vipin Kumar, a recent graduate of the prestigious Indian Institute of Science, who had come to Caltech from an initial postdoctoral appointment at Harvard University, and James L. Urban, a trained pathologist who had by then gotten his Caltech Ph.D. with Professor Hood and moved on to a regular faculty position at the University of Chicago.[1]

Kumar and Urban had started work in the Hood group at about the same time—initially working together, but then independently, in a hot area of immunology having to do with autoimmune diseases such as multiple sclerosis. They were both under intense pressure and worked long hours, drawing praise for their dedication from Hood. Others in the group had in fact accused them of doing "sloppy science,"[2] but when Hood looked into these allegations he declared them to be baseless. Hood, who was a careful scientist, tended to believe that everyone else was equally scrupulous, and he therefore was inclined to discount such assertions.

Hood himself was never accused of misconduct, but in the wake of the allegations he swung into action immediately, sending letters outlining the cases to all those he deemed potentially interested parties, including colleagues, journal editors, and others. He was later criticized for doing that, since the letters told only his side of the story, but he defended his actions. He was understandably anxious that the charges brought against these two junior researchers not otherwise tarnish the reputation of his lab, his coworkers, and the quality and integrity of their science.

Caltech also sprang into action, applying its brand-new rules on research fraud, which, as it happens, I had recently drafted. In June 1990, John Abelson, then the chair of Caltech's Biology Division, initiated an inquiry to determine, as discreetly as possible, whether a probability of fraud existed. It did. Three months later, Provost Paul Jennings appointed a committee of four Caltech biologists, under the leadership of James Strauss, to investigate the allegations in the Kumar case. A smaller group—dubbed "the Fraud Squad" and made up of Jennings, myself, and Caltech lawyer Sandy Pool—was assigned to manage the outreach aspects of the problem, such as notifying the relevant federal agencies, journal editors, and so forth. Even before the Caltech investigations were complete, Hood had retracted three papers by Kumar and Urban.

In the end, the most serious accusation brought against Kumar was that he faked a figure in one of his papers, published in the December 1989 issue of the *Journal of Experimental Medicine* (*JEM*). The figure, a "Southern blot," was supposed to show that DNA from several different cell lines all had the same configuration. However, an analysis of the random spots that showed up in that figure revealed that only a few lines were originally used and were then duplicated to seem like more lines. Although everyone (both the members of Hood's lab

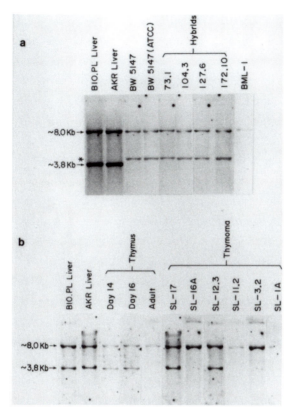

Figure 3.1
The Southern blot. Notice the spots on the upper halves of lines 3, 5, and 7 and the ones on the upper left of lines 4, 6, and 8. On the lower traces, notice the spots on lines 2 and 6 and on 7 and 10. All these are duplicate lines. Photo courtesy of James Strauss.

and the peer reviewers for the journal) missed this in the original review, it becomes vividly obvious in figure 3.1, once it is pointed out.

Kumar reportedly explained that he had just been trying to prepare a more compelling figure and was unaware that such duplication was not acceptable,[3] but everyone who subsequently saw it, especially the members of the Strauss committee, considered it to be a gross misrepresentation. The committee further found that Kumar's transgressions were not limited to that single figure and that he had also made misleading statements in a paper published in the *Proceedings of the National Academy of Sciences (PNAS)*.

Kumar's defense was that he was green and naïve. He said he had never before prepared a paper, that work having previously been done by his supervisors both in India and at Harvard, even though he had appeared on these articles as a coauthor.[4] That argument did not move the investigating committee, which said that it found not credible Dr. Kumar's statement that he did not feel it was wrong to mislabel figures. The committee's report went on to note that Kumar had a Ph.D. degree from the Indian Institute of Science, which is widely regarded as among the best Indian universities, and had been a postdoctoral fellow for two years at Harvard. His recommendations from both universities were excellent. He was not, the report concluded, a naïve, beginning student upon coming to Caltech.

Kumar did have a defender in Eli Sercarz of UCLA, a former collaborator of Hood's, who took Kumar in as a postdoc during his suspension from Caltech.[5] Sercarz argued that for Hood to have written letters to various interested parties essentially incriminating Kumar and exculpating himself even before the Strauss committee report was out was particularly unfair, a denial of due process. However, most observers agreed that Caltech handled the matter in exemplary fashion.

Indeed, after following up on the Caltech investigation and conducting its own independent review, the Office of Research Integrity (ORI, see the next chapter) upheld Caltech's finding

Figure 3.2
Leroy Hood, 1978.
Courtesy of Robert Paz /
Caltech Public Relations.

that Kumar had "committed scientific misconduct by falsifying and/or fabricating figures" in two scientific papers, one of them published and the other submitted for publication to the journal *Cell*, and had "made materially misleading statements in a paper published in the *Proceedings of the National Academy of Sciences*."[6] (The *Cell* paper had been withdrawn prior to publication, and the *JEM* and *PNAS* papers were retracted by Hood after the issue first came to light at Caltech.)

Urban's transgression, which was uncovered by the Strauss committee in the course of the Kumar investigation, was of a different nature. He had produced a paper concerning a mouse model of multiple sclerosis (on which Hood appeared as a co-author) containing phony data, which he submitted to a journal and sent out for peer review. Later, when the deception was brought to light, he maintained that he had planned to obtain the real data and substitute them for the made-up version before

Figure 3.3
James Strauss, ca. 1989.
Courtesy of James Strauss.

the paper was published. He subsequently stated that he felt himself to be under enormous time pressure and that this was his way of handling the situation. He was convinced that he knew how the experiment would come out if he did it properly, so he felt justified in fabricating the initial data and getting the paper reviewed while he was taking the actual data. The paper was published in *Cell* in October 1989, shortly before the discovery that the preliminary data had been falsified.[7]

Once again, Caltech's Biology Division chair conducted a preliminary inquiry and convened a committee consisting of the same four biologists who had investigated Kumar, along with a participant from outside the division. The committee found that the data in the published version, although they may have been accurate, were not the same as those in the manuscript that had been submitted for review. This was still in the early days of the research-misconduct business, and Caltech drew a distinction between research misconduct, which it concluded had occurred, and outright research fraud (involving deliberate intent to deceive), which Caltech believed had not happened in the case of either Urban or Kumar. That distinction is no longer considered important either at the institute or within the larger scientific community.

In May 1991, Kumar was fired from his postdoctoral position, the most severe penalty that the institute was capable of applying. Urban was no longer at Caltech when the investigation into his affair was completed, so he received only a letter of reprimand. However, the University of Chicago, which had hired him as an assistant professor, was informed of the entire affair, and he agreed to resign his position there. Thus both transgressors were at least temporarily removed from the world of science.

As of this writing, Urban seems to have faded altogether from the scientific scene. Kumar, by contrast, appears to have landed on his feet. He served out a three-year banishment from National Institutes of Health (NIH) funding and then, after stints at UCLA and the La Jolla Institute for Allergy and Immunology, became the head of the Laboratory of Autoimmunity at the Torrey Pines Institute for Molecular Studies in La Jolla, California. More recently, he has become a researcher at the Multiple Sclerosis National Research Institute in San Diego. It is to be hoped that he has permanently learned his lesson. However, more careful supervision at all stages of his education might have avoided the problem in the first place.

ᴵᶠ Four
Codifying Misconduct:
Evolving Approaches in the 1990s

In the late 1970s, matters of scientific integrity (and the occasional lack thereof) in the nation's research laboratories began to capture the attention of the American public. In 1981, future vice president and then Tennessee congressman Albert Gore, Jr., chairman of the Investigations and Oversight Subcommittee of the House Science and Technology Committee, held the first hearings on the emerging problem. In 1985, Congress enacted the Health Research Extension Act, which required institutions seeking research funds from the Public Health Service (PHS), the oversight body of the NIH, to establish "an administrative process to review reports of scientific fraud" and "report to the Secretary any investigation of alleged scientific fraud which appears substantial." Four years later, in March 1989, the Office of Scientific Integrity (OSI) was created under the jurisdiction of the director of the National Institutes of Health (NIH). The name "Office of Scientific Integrity" was originally proposed as a kind of Orwellian joke, but it stuck, at least for the short life of the new agency.

The OSI mandate was to protect the integrity of scientific research in the United States by investigating allegations of scientific fraud—largely in biology, a natural focus for an agency administered by the NIH—that came its way. In May 1992, after encountering some difficulties in getting its decisions to stick, the OSI and its oversight body, the Office of Scientific Integrity Review (which had been created at the same time), would be moved out of the NIH by the secretary of the Department of Health and Human Services and combined to form the Office of Research Integrity (ORI), directly under PHS oversight. At about the same time, the National Science Foundation (NSF) created an Office of the Inspector General (OIG), which would investigate allegations of misconduct—the government disliked using the term "fraud" in its official language—that occurred in the course of research investigations funded by the NSF. The first set of regulations, issued on July 1, 1987, defined misconduct as "fabrication, falsification, plagiarism, or other serious deviation from accepted practices in proposing, carrying out, or reporting results from research."[1] Although the agency's definition also addressed "the material failure to comply with federal requirements for protection of researchers, human subjects, or the public or for ensuring the welfare of laboratory animals," and "failure to meet other material legal requirements governing research," the NSF seemed most interested in the first item.

Both the ORI and the OIG soon ran into difficulties, largely because this definition, so broadly conceived as to cover every possible contingency, proved to be of limited practical use. In October 1991, OSI deputy director Suzanne Hadley gave a speech at the University of California, San Diego in which she noted: "It is essential that the observation, data recording, and data interpretation and reporting be veridical with the phenomena of interest [and] be as free as humanly possible of the 'taint' due to the scientists' hopes, beliefs, ambitions, or desires." I have already noted that such an inductivist approach is not necessarily good for science. In the same speech, Hadley continued, "The really tough cases to deal with are the cases closest to the average scientist: those in which fraud"—Hadley did not hesitate to use the term in public—"is not clearly evident but 'out of bounds' conduct is: data selection, failure to report discrepant data, overinterpretation of data."[2]

The idea that data selection and overinterpretation of data inherently constitute forms of misconduct arises naturally out of Hadley's inductivist (that is, Baconian) view of the scientific method, but condemning such marginal research methodologies out of hand is not good scientific practice. Hadley, incidentally, was trained as a psychologist, not as a typical bench scientist.

The pitfalls of this approach became apparent as soon as the OSI and its successor agency, the ORI, undertook to investigate allegations of misconduct in the labs of two high-profile scientists, the renowned virologists Robert Gallo and David Baltimore. The details of these scientific *causes célèbres* are well known.[3] In both cases, the scientists' colleagues—Mikulas Popovic in Gallo's lab, Thereza Imanishi-Kari in Baltimore's—were accused of misconduct in a spiraling cycle of charges, investigations, and legal actions that ultimately grew to involve Representative John Dingell (D.-Mich.), the powerful and publicity-savvy chair of the House Subcommittee on Oversight and Investigations of the Committee on Energy and Commerce. In the Baltimore case, Imanishi-Kari's postdoc Margot O'Toole accused her of falsifying immunological data in a 1986 article that Imanishi-Kari, Baltimore, and others had published in the journal *Cell*.[4] O'Toole brought the charge after failing to replicate the original findings in her own experiments. Her allegations were subsequently taken up by two staffers at the National Institutes of Health and then by Dingell's subcommittee. In the resulting furor, Baltimore, who had reluctantly agreed to retract the *Cell*

Figure 4.1
David Baltimore, 1994. Courtesy
of Donna Coveney, M.I.T. / California Institute of Technology
Archives.

paper while vehemently defending Imanishi-Kari's integrity as
a scientist and was never himself charged with any wrongdoing,
was forced out of his job as the newly appointed president of
Rockefeller University. Imanishi-Kari, an assistant professor at
Tufts University, had her tenure decision postponed and her
research and research methodologies subjected to many months
of intense OSI and Secret Service scrutiny.

The Gallo case, which unfolded in the glare of international
media coverage, centered on whether Gallo's lab at the National
Institutes of Health had wrongly claimed credit for discovering
the virus that causes AIDS when the key breakthrough had really been made by Luc Montagnier and his team of scientists at
the Pasteur Institute in Paris. This time it was Popovic who was
at the center of the storm—for allegedly misappropriating cell
cultures that had been supplied to the lab by Montagnier's team
at the Pasteur. Once again, the case wound up in the hands of
the OSI, whose investigators found Popovic guilty of "misrepre-

sentations or falsifications of the actual methodology and data."[5]
A year later, its successor agency, the ORI, found Imanishi-Kari
guilty of multiple counts of research misconduct and recommended that she be barred from receiving federal research funds
for a period of ten years.

Both these decisions were subsequently overturned by
Health and Human Services appeals boards, which found that
the OSI and ORI had made errors that rendered their verdicts
meaningless. In 1997, Baltimore became president of Caltech,
a position he held for the next nine years; he remains at the
institute as the Robert Andrews Millikan Professor of Biology.

Figure 4.2. Robert Gallo, ca. 1981. Courtesy of NIH.

Imanishi-Kari was granted tenure at Tufts and is today an associate professor of pathology there; Gallo is currently the director of the Institute for Human Virology of the University of Maryland Biotechnology Institute, where Popovic is employed as a professor and researcher. In 2008, Montagnier and his Pasteur colleague Françoise Barré-Sinoussi shared half of the Nobel Prize in physiology or medicine "for their discovery of human immunodeficiency virus," or HIV, the virus that causes AIDS. Gallo did not share in the award.

The heated controversies and near-Byzantine complexities of the Gallo and Baltimore cases have been amply and exhaustively documented in numerous articles and several books, and it is not my intention to delve into the details here. Suffice it to say that both cases cast a less than flattering light on the federal government's initial efforts to grapple with scientific misconduct. Some hint of these difficulties had already surfaced in 1991, when, in an effort to codify their positions, the OSI and the OIG issued nearly identical definitions of scientific misconduct in the *Federal Register,* the official daily publication for rules, proposed rules, and notices of federal agencies and organizations, as well as executive orders and other presidential documents. These stated that "Scientific misconduct is . . . fabrication, falsification, plagiarism, or other practices that seriously deviate from those that are commonly accepted within the scientific community."[6]

That same scientific community found itself in general agreement with the "fabrication, falsification [or] plagiarism" part (which soon acquired the acronym "ffp," or, as an anonymous wag dubbed it, "frequent-flier plan"). However, the rest of the definition, regarding "practices that seriously deviate from those that are commonly accepted," brought forth waves of protest. There are many practices that are not commonly accepted within the scientific community but don't, or shouldn't, amount to scientific fraud.

In 1988, I became Caltech's vice provost, and it came to my attention from various pieces of paper that crossed my desk that the institute, like every other research university, would soon need to have a formal policy in place, in the event a case of scientific misconduct arose. Accordingly, I read everything about the topic I could get my hands on and drafted a policy. After surviving the scrutiny of the various committees that run Caltech, it was ultimately adopted and became the law of the land.

Even before the policy became official, it got tested. A Nobel laureate from another institution wrote to the director of the NSF alleging that one of our professors had committed fraud. Caltech's president at the time, Thomas E. Everhart, decided that even though my rules hadn't yet been adopted, at least they had been drafted and ought to be applied. The rules called for the first phase of such cases to be an inquiry, carried out with the maximum possible discretion, into whether fraud had actually occurred. The inquiry was supposed to be conducted by the chair of the accused faculty member's division (Caltech is organized into six academic divisions). In this case, however, the chair knew relatively little about the field in question, which was pretty close to my own. So I wound up doing the inquiry. After a solid month of hard work, I decided that our professor was guilty of nothing more than an excess of enthusiasm, but I took the precaution of having him withdraw and rewrite some papers that had not yet been published. That procedure satisfied everyone. Shortly thereafter, my proposed regulations governing scientific misconduct at Caltech were ratified, and they have remained in place, with some minor modifications, ever since. (Caltech's policy on scientific misconduct, in its present form, is included in this book as an appendix.)

Compared to the federal regulations in place at that time, which the government regarded as binding on all universities including Caltech, our new rules were considerably narrower in defining what constituted fraud. Nor was the distinction between the Caltech definitions and the OSI/ORI standards purely academic. Twice in the early 1990s, young faculty members who were up for tenure at the institute were accused by external referees (outside academics from whom the tenure committee had requested references) of committing scientific fraud. When a thing like that happens, tenure review committees typically go bananas. They have no idea how to respond to such accusations. In both these cases, the chair of the committee came to the local expert (me) to ask for advice.

The first case involved a professor of physics who was working in an intensely competitive field of research. Before coming to Caltech, while working at a different institution, he had put down the names of two people as coauthors on one of his papers without their consent. The second case concerned a faculty member in biology who had published what appeared to be the same paper in two prestigious journals. The editors of both journals were very upset. In actual fact, the professor had achieved a breakthrough in the months between the first and second publication and had found it necessary to recapitulate a substantial portion of the original paper in the subsequent publication, in order to put the breakthrough in context.

In both cases, I was able to assure the respective tenure committees that, according to the Caltech definitions, neither young researcher had engaged in conduct that amounted to scientific fraud. However, it was certainly the case that according to the government's definition, with its "deviate from . . . commonly accepted" clause, the actions might well have qualified. Both young people were promoted to tenured positions and are today among Caltech's brightest stars.

During this period, I received a letter from an ORI official informing me that Caltech's rules had been examined and deemed inadequate because of their lack of compliance with federal regulations. I was further advised that we should change the Caltech rules and notify ORI of the change within ninety days. Caltech could lose all its government funding if we didn't comply, a very serious matter. Eighty-nine days later, I wrote back, saying that it took time to change rules in a university and that the change had to be approved by committees and so on. The official obligingly wrote me back saying that this was all right as long as I let him know once the change had been made. Our lawyers interpreted this "extension" to mean that we had another ninety days to comply, but I interpreted it to mean that we had no deadline at all, and I did not act on his request. As things turned out, the government's wording would change before all of this had any consequences for Caltech.

The general displeasure of the scientific community with the existing rules prompted the federal government to convene in April 1996 yet another blue-ribbon committee of scientists to consider the matter, this time by the office of the president's science advisor. In 1999, after much work, this panel did come up with a suitable definition of scientific misconduct with which most scientists could agree. The new language retained the ffp, "fabrication" being defined as making up results, "falsification" as changing or omitting data or results, and "plagiarism" as the appropriation of ideas without credit. But the new standards also required that the conduct in question constitute "(1) a significant departure from accepted practices of the scientific community for maintaining the integrity of the scientific record; (2) The misconduct be committed intentionally, or knowingly, or recklessly disregarding accepted practices; and (3) The allegation be proved by a preponderance of evidence."[7] That October, a request for public comment on the proposed federal research misconduct policy appeared in the *Federal Register*, which drew more than two hundred comments from interested parties. A year later, on December 6, 2000, the new federal policy on research misconduct went into effect.[8]

In other words, the new regulations downgraded the old OSI's "practices that seriously deviate from those that are commonly accepted within the scientific community" to secondary status—as an additional requirement, rather than retaining it as a sufficient, primary cause for a finding of scientific misconduct. This was a definition that Caltech could live with. Within weeks after the new government regulations came out, the institute adopted their exact words, which were in any case equivalent to Caltech's original language, and it is now fully in compliance with the federal rules.

⌐ Five
The Cold Fusion Chronicles

On December 6–9, 1993, the Fourth International Conference on Cold Fusion took place in Hawaii, on the island of Maui. The event had all the trappings of a normal scientific meeting. At least 250 scientists took part, mostly from the United States and Japan (hence the site in Hawaii), with a sprinkling from Italy, France, Russia, China, and other countries. More than 150 scientific papers were presented, on such subjects including calorimetry (a measurement of how much something warms up when you put a given amount of heat into it), nuclear theory, materials, and so on. The founders of the field, Stanley Pons and Martin Fleischmann, were in attendance and were treated with the deference due their celebrity status. At the time, they were carrying out their research in a laboratory that Technova, a subsidiary of Toyota, had built especially for them in Nice, on the French Riviera. At the meeting, it was announced that the Japanese trade ministry, MITI, had committed $30 million over a period of four years to support research on what was delicately called "new hydrogen energy," including cold fusion.

Contrary to appearances, however, this was no normal scientific conference. Cold fusion had become a pariah field, cast out by the scientific establishment. Between its practitioners and "respectable" science there was virtually no communication. Cold fusion papers were almost never published in refereed scientific journals, with the result that those articles didn't receive the normal critical scrutiny that science requires. At the same time, because the cold fusion contingent saw itself as a community under siege, there was little internal criticism. Experiments and theories tended to be accepted at face value, for fear of providing even more fodder for external naysayers, if anyone outside the group was bothering to listen. In these kinds of circumstances, crackpots flourished, and this made matters worse for those who were at least willing to entertain the notion that there might have been serious science going on.

Here it is important to draw the distinction between scientific fraud and what Irving Langmuir called "pathological science."[1] Although there is no evidence of scientific fraud in the cold fusion story, it does come rather close to what Langmuir defined as pathology. In pathological science the person involved always thinks he or she is doing the right thing, but is led into folly by self-delusion. Examples include the celebrated case of N-rays and various other kinds of rays "discovered" in the wake of X-rays, as well as J. B. Rhine's extrasensory perception, and the many reported cases of flying saucers. Those all share the distinction of being wrong. Aspects of cold fusion, on the other hand, may well turn out to have been right. We don't know yet.

In its brief hectic heyday, cold fusion's rise and fall was widely and vociferously documented in the press and popular literature. The furor began on March 23, 1989, when Pons and Fleischmann, fearing they were about to be scooped by a competitor named Steven Jones from nearby Brigham Young University, and with the encouragement of their own administration at the University of Utah, held a press conference. There they announced that, incredible as it sounded, they had

Figure 5.1
Press conference with Martin Fleischmann, left, and B. Stanley Pons,
May 1989. Courtesy of AP Images.

induced controlled nuclear fusion reactions on a bench in their laboratory. If true, their discovery undeniably ranked as one of the chief scientific breakthroughs of the twentieth century.

At the time that Pons and Fleischmann made their announcement, hot fusion—the fusion of two atomic nuclei into one, with a concomitant release of considerable energy—was already an old story. It's real all right, since it takes place inside the sun and all the stars, but making it happen here on Earth is another matter. It is the phenomenon that powers hydrogen bombs, but this type of uncontrolled, runaway reaction is of no use as a safe, reliable energy source. *Controlled* hot fusion is what's wanted, particularly since, unlike controlled nuclear fission (which involves splitting atomic nuclei rather than fusing them), it does not produce radioactive waste as a by-product. The achievement of controlled hot fusion, which involves confining a very hot dense gas, or plasma, of ionized deuterium (heavy hydrogen) atoms in an effort to induce the nuclei to fuse, has been twenty-five years away for the past sixty years. If past results are anything to go by, it probably still is twenty-five years away. However, if hot fusion could be reliably brought under control in the laboratory, it would solve all our energy problems, since every gallon of seawater contains enough deuterium to yield the energy equivalent of 300 gallons of gasoline. (Deuterium is a hydrogen isotope whose nucleus contains not only a proton but also a neutron—hence the term "heavy hydrogen.")

What Pons and Fleischmann claimed to have discovered was an inexpensive, essentially low-tech way to make fusion occur.[2] Their method did not require teams of scientists, plasma technology, and mega-levels of research funding. Rather, they said that they had achieved their results through the relatively humble means of electrolyzing, or passing an electric current through, heavy water (which is formed when deuterium bonds with oxygen, creating D_2O rather than our familiar H_2O). The costliest part of the entire enterprise came from the electrodes, made of the precious metals palladium and platinum. If these claims were to be believed, the world's energy problems were at an end, to say nothing of the fiscal difficulties of the University of Utah.

What followed the chemists' announcement was a kind of feeding frenzy—science by press conference and e-mail, confirmations and disconfirmations, claims and retractions, ugly charges and obfuscation. In short, it was science gone berserk. Then, as abruptly as the tempest blew in, it was gone. For all practical purposes, it ended a mere five weeks after it began, on May 1, 1989, at a dramatic session of the American Physical Society in Baltimore. Although there were numerous presentations at this session, two in particular stood out. Three scientists

Figure 5.2
The three Caltech scientists who challenged Pons and Fleischmann's scientific discovery of cold fusion (left to right): Nathan Lewis, Steven Koonin, Charles Barnes, 1994. Courtesy of Robert Paz/Caltech Public Relations.

from Caltech—physicists Steven Koonin and Charles Barnes in one presentation and chemist Nathan Lewis in the other— executed a perfect slam-dunk that cast cold fusion right out of the arena of mainstream science.[3] In particular, Lewis repeated the experiment and it didn't work for him, and Koonin eloquently described the nuclear theory that made the Pons and Fleischmann result seem impossible. It is also worth noting that this whole exercise added nothing at all to the already sterling reputations of Koonin, Lewis, and Barnes.

Before I go any further with this tale, I will come clean about my own prejudices. The Caltech protagonists—Koonin, Lewis, and Barnes—were at the time faculty colleagues of mine, as well as my personal friends of many years. (Koonin later left Caltech and became the chief scientist for BP. He has since become the United States Undersecretary of Energy.) At the same time, there was on the other side of the game a player who is not only one of my oldest personal friends but also my longtime scientific collaborator. Although his story unfolded largely off the radar screen of journalists and popular authors in the United States, it clearly shows that the commotion that began in Utah was not an isolated or unique phenomenon.

My friend, Professor Francesco (Franco) Scaramuzzi, was at the time the head of a small low-temperature physics research group at a national laboratory in Frascati, a suburb of Rome. This laboratory was run by an agency called ENEA—the National Agency for New Technologies, Energy, and the Environment, roughly analogous to our Department of Energy. It was possible within this agency for a scientist like Franco to be promoted to the rank of *dirigente*, roughly "executive" or "director," a promotion that would not change a researcher's assignment or responsibilities in any substantial way but would carry with it very substantial financial rewards and much prestige. Although Franco was certainly one of the laboratory's more eminent scientists, he had not been awarded this distinction by 1989, when he was sixty-one years old. The reason is that in the corrupt Italian system that then existed (it has since been cleaned up), these promotions were based more on political affiliation than on scientific accomplishment. For every two Christian Democrats promoted, a new Socialist, a Communist, and someone from one of the smaller parties would also be named to the *dirigente*

Figure 5.3
Francesco Scaramuzzi reporting on his discovery of a new kind of cold fusion, Frascati, April 17, 1989. Courtesy of ENEA, Centro Ricerche Energia Frascati.

ranks. Franco had not been promoted because he refused to join a political party in order to advance his professional career as a scientist. He was and to this day remains a man of unflinching integrity.

On the morning of April 18, about a month after the cold fusion story broke in Utah, Franco called me from Rome to warn me that the following day I would find his picture in the *New York Times* (I did).[4] He made the call just after coming out of a press conference in which he had announced his discovery of a new kind of cold fusion. Like scientists everywhere, he had heard the thrilling details of the Utah experiments and decided to try reproducing them himself. He reasoned that electrolysis wasn't really necessary. It served only to get deuterium to insert itself into the atomic structure of the palladium electrode. He also thought it necessary that the experimental apparatus not be in thermodynamic equilibrium (in other words, not be at the same temperature) with its surroundings. He and his handful of young scientists and technicians arranged to put some titanium shavings in a cell pressurized with deuterium gas (titanium is both cheaper and easier to get hold of than palladium, and, like palladium, it is a metal that absorbs large quantities of hydrogen or deuterium into its atomic crystal structure). Then they used some liquid nitrogen (a refrigerant readily available in any low-temperature physics laboratory) to run the temperature of the cell up and down, thus creating thermodynamic disequilibrium.

The Frascati lab's crude apparatus was not suitable for the difficult measurement needed to confirm Pons and Fleischmann's claims that their fusion experiments had generated heat. However, fusion should also produce neutrons (that is what Steven Jones had claimed to have detected at BYU). Franco's team got a colleague to set up a neutron detector near their apparatus and they began running their experiments. During the following weeks, their detector often registered nothing, but on a couple of occasions it indicated very substantial bursts of neutrons. When the experimenters got their second positive result, on April 17, Franco decided he had to inform the head of his laboratory. In no time at all, he found himself in downtown Rome, talking to the head of the entire national agency.

At this juncture, tabletop science ran headlong into big-top politics. For four months, ENEA had been without funding, the necessary legislation having been stalled in Parliament. To meet its payroll, the nation's premier energy agency was borrowing money from banks. All purchases were frozen. Research was paralyzed. To the politically astute ENEA head, Franco's discovery was too good an opportunity to pass up. Franco agreed to a press conference, but only if he could first give a full technical seminar to his scientific peers. His presentation, hastily organized for that same day, was crammed to the rafters with scientists from every laboratory in the Rome area and was even covered by the evening television news programs. At the press conference the next morning, Franco was stunned to find himself flanked by two ministers of state. He recapitulated his findings with the utmost scientific objectivity and restraint, but his measured language and dignified deportment made not the slightest bit of difference. The story made headline news all over Italy. Within days, the Italian parliament had approved financing for ENEA and Franco had been promoted to *dirigente*.

He was also being hailed as the Italian Prometheus, stealing fire from the gods. My reserved, correct, self-effacing friend had become a media celebrity, suddenly the most famous scientist in Italy. When I visited him in Frascati just a few months later, in the summer of 1989, he handed me two binders, each two or three inches thick, full of photocopies of his press notices in Italy and abroad. Although these events in Italy escaped the notice of most Americans, what happened there had mirrored in many important ways the frenzy in the United States.

For one thing, in both cases, pecuniary motives had driven science out of the confines of the laboratory into the blinding glare of publicity. For another, the story instantly captured the public fancy. Not only were the gallant scientists about to rescue us from the clutches of the greedy oil barons (the whole affair took place not long after the *Exxon Valdez* incident), but also the saga was spiced with delicious ironies. In the United States, mere chemists, spending money out of their own pockets, had apparently succeeded where arrogant physicists, spending hundreds of millions in public funds, had conspicuously failed. The chemists had outshone the physicists, small science had beaten big science, sheer ingenuity had triumphed over brute force, and humble professors from Utah had won out over the scientific elite of bicoastal, non-Mormon America. Somehow lost in this modern-day David and Goliath fable were the facts that two of the Utah professors, Pons and Jones, were bitter rivals; Jones, the only Mormon of the bunch, was a physicist, not a chemist; and Pons' partner, Fleischmann, was not only an Englishman but a fellow of Britain's Royal Society, one of the world's oldest and most illustrious scientific bodies. These were mere quibbles, however.

In Italy, the story had unfolded along remarkably similar lines. There the dire straits of ENEA drove the story out of the lab and into the headlines. Not only had cold fusion been reproduced in Italy, the Italian version was of an entirely new kind: *Fusione fredda*, or cold fusion Italian Style, was "dry fusion," produced without electrolysis. True, Franco was also a physicist, not a chemist, but his specialty was not the "big science" research for which the Frascati lab was famous but instead small, clever, low-budget science. Suddenly, Italy had more to offer the world than sunshine, pasta, and the ultimate Renaissance immersion experience. For perhaps the first time since the days of Enrico Fermi, an Italian scientific hero strode the world stage like a colossus (or so it seemed from inside Italy).

With the advantage of twenty years' hindsight, we can see today how thoroughly the cold fusion episode seemed to stand science on its head. This was not only because it played out in the popular press without the ritual of peer review but also because both sides of the debate violated at least two of the prevailing canons of scientific logic. As I noted in the first chapter of this volume, science in the twentieth century has been much influenced by the falsifiability precepts of the Austrian philosopher Karl Popper, which hold that while a scientific idea can never be conclusively proven true, a single contrary experiment may suffice to prove a theory forever false. In this formulation, science advances only by demonstrating that theories are false, at which point they must be replaced by better ones. However, as I also noted, Popperian doctrine in its purest form offers an unrealistic and somewhat misleading model for scientific conduct, since even the most dedicated scientists can hardly be expected to question the validity of their research with the same zeal that they bring to pursuing it (although there are some who do). But there is no question that inasmuch as it refers to self-correcting mechanisms within the scientific community as a whole, falsifiability aptly describes key aspects of how science functions. What distinguished the proponents of cold fusion in this regard was the readiness with which they embraced precisely the opposite view when many experiments, including their own, failed to yield the desired results. These experiments were irrelevant, they argued, incompetently done, or lacking some crucial (perhaps unknown) ingredient needed to make the thing work. Only the positive results—the appearance of excess heat or a few neutrons—were to be credited, and these proved that the phenomenon was real. This anti-Popperian flavor of cold fusion played no small role in its downfall, since seasoned experimentalists like Caltech's Nathan Lewis and Charles Barnes refused to believe what they couldn't reproduce in their own laboratories.

On the other hand, the anti-cold fusion crowd was equally guilty of violating another of the solemn scientific canons that we adduced (and qualified) earlier—namely, that science must be firmly rooted in experiment or observation, unencumbered by preexisting theoretical baggage. And yet a key element in the downfall of cold fusion was that it claimed experimental results that ran contrary to prevailing theory.

All parties agreed that if cold fusion had occurred in the experiments of Pons and Fleischmann, Jones, Scaramuzzi, and many others, the primary event would have to have been the fusion of two deuterium nuclei. This in itself would have been a reasonably rare event, because deuterium's positively charged nuclei, like two north (or south) magnetic poles, repel each other. However, if the nuclei get close enough together, they will nevertheless fuse because of what is called the strong nuclear force. The laws of quantum mechanics, which govern the behavior of matter at the atomic scale, also allow deuterium nuclei to fuse by accident every so often, even if they are not initially close together—but the probability of that happening is very small. When the nuclei are as far apart as they normally are in a deuterium molecule, the probability that they will "happen" to fuse is much too small to have produced the alleged effects claimed by the cold fusion crowd.

There are two ways to look at just how small the probability is. At the inter-nuclear spacing in the deuterium molecule, the probability is too small by forty or fifty orders of magnitude. As one order of magnitude means a factor of ten, to say that something is too small by forty or fifty orders of magnitude really means that it is too small beyond discussion, beyond imagination, almost beyond meaning. On the other hand, that probability is insanely sensitive to how far apart the nuclei are to begin with. To increase the probability of fusion by the requisite forty or fifty orders of magnitude requires getting the nuclei closer together by just one order of magnitude. While this may sound like a trivial challenge compared to bridging forty to fifty orders of magnitude, it is in fact extremely difficult to pull off. Hundreds of millions of dollars have been spent on hot fusion research with the goal of producing exactly that elusive result, and it is difficult to imagine how—given the well-known forces involved—it could be accomplished on a tabletop. Nevertheless, once we have been anesthetized by talking about forty or fifty orders of magnitude, the idea that a one-order-of-magnitude gap might somehow be overcome is not so hard to swallow.

But the theoretical difficulties of cold fusion didn't end with getting deuterium nuclei somehow to fuse. When two deuterium nuclei do fuse, they momentarily form the nucleus of a common isotope of helium, called helium-4. However, that fusion reaction produces so much excess energy that the helium-4 nucleus almost always breaks up immediately into two smaller pieces. About half the time, a neutron pops out, leaving a helium-3 nucleus. The other half of the time, a proton peels off, leaving a hydrogen-3, also known as tritium, nucleus. It also happens that about one time in a million, the helium-4 nucleus doesn't break up at all. Instead it zooms off still intact, while emitting a powerful gamma-ray photon. In all three cases, the two pieces take off in opposite directions, carrying lots of energy.

What we expect, then, is that about half the deuterium fusions will produce energetic neutrons, while the other half will leave behind tritium as evidence that they occurred. In fact, as we have already seen, Jones, Scaramuzzi, and others did detect neutrons, which they interpreted as evidence for cold fusion. However, there were always far too few neutrons to account for the amount of heat that Pons and Fleischmann claimed to have measured. In fact, on the evening of the original Pons and Fleischmann press conference, I ran into one of my buddies at Caltech, a battle-scarred veteran of experimental nuclear physics. "What do you think?" I asked (there was no need to be more specific). "It's bullshit," he said, slipping immediately into technical jargon. "If it were true, they'd both be dead." What he meant was that if enough fusion reactions had taken place to produce the amount of heat claimed by Pons and Fleischmann, the associated flux of neutrons would have been more than enough to send them both to the happy hunting grounds.

To believe that Pons and Fleischmann, Jones, Scaramuzzi, and many others who claimed to have observed heat, neutrons, or tritium were all observing the same phenomenon would require us to believe that when fusion occurs inside a piece of metal such as palladium or titanium, the outcome is radically different from what is known to happen when fusion occurs in the sun, a hot fusion plasma, a hydrogen bomb, or a nuclear accelerator. It is, in other words, an outcome that contravenes all that we know about conventional nuclear physics. Let us catalog three possible outcomes of fusion: one that emits neutrons (a); one that leaves tritium behind (b); and the rare event where the helium-4 stays intact (c). In conventional nuclear physics, fusion results about half the time in a, half the time in b, and one millionth of the time in c. To account for the observations reported, with some consistency, by various cold fusion researchers, fusion inside a metal would nearly always result in reaction c (without, however, emitting a gamma ray). One in every 100,000 or so reactions would result in b, and the probability of a reaction a would be an additional 100,000 times smaller. These are the conditions needed to explain why cold fusion cells could allegedly generate enough energy to light a light bulb for periods of days or months, while yielding only tiny amounts of tritium, and barely detectable traces of neutrons that fell far short of killing Pons and Fleischmann.

Is it plausible that nuclear reactions might be altered radically when they occur among the atoms in a metal rather than in the rarefied atmosphere of hot plasma? The answer, quite simply, is no. For one thing, compared to the distances between atoms in a metal, the atomic nucleus is so small that for all practical purposes it always occupies a near vacuum. For another thing, events occur so rapidly in the fusion reaction that metal is simply unable to respond. If you like orders of magnitude, the fastest that anything can happen in a metallic crystal is nine orders of magnitude (that is, 10^9 times) slower than the typical time it takes for two deuterium nuclei to fuse into a single entity and break up. In other words, compared to the lightning speed with which the nucleus does its thing, the atoms of the crystal are far away and frozen in time. Finally, the energy released in the nuclear reaction is so large that the metallic crystal has no means to absorb it unless that energy spreads out instantaneously, over vast distances, by some unknown mechanism. (Presumably, the same mechanism would have to account for the absence of any gamma ray emissions.) In short, according to everything we know about the behavior of matter and nuclei, cold fusion cannot exist.

However, as I noted in chapter 1, if the history of scientific discovery has taught us anything, it is that nature will from time to time confound us with behavior that everyone previously "knew" to be impossible. Back in 1989, there were two examples that seemed particularly relevant to cold fusion: high-temperature superconductivity and the Mössbauer effect.

In 1986, two Swiss physicists, J. Georg Bednorz and Karl A. Müller, announced the discovery of a ceramic-type material that remained superconducting at temperatures as high as 30 kelvins (K) (that is to say, −243°C or nearly −400°F).[5] They were already working in a field whose manifestations violate the trained intuition of physicists, superconductivity being a phenomenon in which many metals, cooled to sufficiently low temperatures, are able to conduct electricity with absolutely no resistance while simultaneously expelling completely any applied magnetic field. This behavior is so bizarre that it took nearly half a century after its discovery in 1911 to formulate an acceptable theoretical explanation. There was a certain consolation, however, in realizing that if nature insisted on playing such weird tricks, at least they were confined to the privacy of the physics laboratory by the requirement of extremely low temperatures. Before Bednorz and Müller made their discovery, it was well known that superconductivity could "never" exist at a temperature higher than 35 K, but Bednorz and Müller had nearly overcome that barrier. Afterward, it was a mere matter of months before the discovery of materials that remained superconducting up to 100 K. That's still pretty cold—normal room temperature is about 300 K—but the stunning impact of that discovery on the scientific community is hard to overstate, and contributed to the cautiously receptive atmosphere that greeted the initial announcement of cold fusion. (The events surrounding the discovery of high-temperature superconductivity and its aftermath are the subject of my final chapter.)

The Mössbauer effect, discovered thirty years earlier, in 1957, was another completely unexpected occurrence that seemed to have an even more direct bearing on cold fusion. We have already seen that one reason cold fusion is hard to credit is because of the implausible idea that a nuclear reaction might be altered in any meaningful way simply because it takes place in a crystal. Yet the Mössbauer effect—named for its discoverer, German physicist Rudolf Mössbauer, who received the Nobel Prize in 1961 as a result—was a phenomenon in which precisely that does seem to happen.

When a nucleus has too much energy, it must find some means to get rid of the excess. For example, we've already seen that when two deuterium nuclei fuse, the extremely energetic nucleus that results can actually break up in any of three ways. In all three cases, however, the result is two fragments that fly off in opposite directions. Mössbauer's discovery was that in certain cases when a nucleus in a crystal gives up its excess energy by emitting a gamma-ray photon, the two fragments do not invariably go their opposite ways.[6] Instead, there is a substantial probability that the photon will fly off while the nucleus remains stationary. That is because the force of the recoil that would under other circumstances send the nucleus flying away is absorbed by the entire crystal, leaving the nucleus essentially motionless. The net result is that the gamma-ray photon emitted by a nucleus in a crystal can have slightly more energy than the gamma-ray photon emitted by the same nucleus in a vacuum. Our carefully trained intuition—which says that a crystalline environment can have no possible bearing on the behavior of atomic nuclei because the two entities exist in entirely separate realms of distance, time, and energy—has been violated. If our intuition could be contradicted by the Mössbauer effect, then why couldn't the same argument be made for cold fusion?

That's a good question, and there are very good answers. First, the Mössbauer effect can be observed only for a few special nuclear reactions in which the initial excess energy is much smaller, and the time the nucleus takes to get rid of it much larger, than in the cold fusion reaction. In other words, it occurs precisely in those special cases where our assumption that the nucleus and the crystal act on incompatible scales of time, distance, and energy no longer holds true. Second, even then, the Mössbauer effect does not change the intimate details of the nuclear reaction, such as whether it produces a gamma-ray photon, or the probabilities of the various possible ways (our *a*, *b*, and *c*) that the excess energy is released. As we have seen, it is precisely these aspects of the reaction that must be radically different if cold fusion is real. Finally, the Mössbauer effect is in a sense the exact opposite of what is supposed to happen in cold fusion: Instead of the nuclear recoil energy somehow turning into heat in the atomic lattice (as was said to be the case in cold fusion), the Mössbauer effect is the special case in which no heat at all is produced. It is interesting precisely for this reason.

Nevertheless, in spite of all the differences, many scientists instantly thought of the Mössbauer effect when they first heard of cold fusion. Mössbauer's discovery had been unexpected, but once it happened it was quickly and satisfactorily explained within the framework of conventional theory. It reaffirmed that there are still genuine surprises waiting for us that, once understood, don't violate conventional physical laws. It also proved that there are at least some realms in which nuclear physics and solid-state physics affect each other.

All of this helps to explain why, immediately after the March 1989 press conference in Utah, most scientists were willing at least to give cold fusion a chance. It was precisely during this crucial probationary period (so to speak) that cold fusion science went, as scientists are wont to say, nonlinear. Many scientists tried their own hand at it. Those who succeeded, or who seemed to succeed, in replicating the Utah results held press conferences. Those who failed generally let the matter quietly drop and went on to other things. It would be difficult to devise a worse way of doing science. Among the exceptions to that behavior were Lewis, Barnes, and Koonin of Caltech. They pursued every avenue with relentless tenacity and Popperian rigor, repeating every experiment, calculating every effect, looking not merely for positive or negative results but also for explanations of the ostensibly positive results that others were reporting. These they found in abundance. Far from publicizing their work, they initially kept their findings so quiet that rumors started to circulate, and even to appear in the press, that they were protecting positive results. Finally, they were able, five weeks after the Utah press conference, to stand before their colleagues in Baltimore and, piece by piece, in vivid detail, demolish the case for cold fusion. Cold fusion had been given its chance and granted a suspension of disbelief, no matter how unlikely the claims for it seemed, and it had failed to prove itself. In the eyes of respectable science, cold fusion was essentially dead.

There were a few notable deviations. Back in Frascati, Franco Scaramuzzi and his group of young researchers were not quite prepared to give up just because halfway around the world a trio of scientists from southern California didn't approve. Franco himself had experienced not just fifteen minutes of fame but a month of it, and the attention showed no signs of letting up. He was a hero, not only to the Italian public but also to all his colleagues in ENEA, and ENEA itself had suddenly shed its reputation for bumbling bureaucratic ineptitude. This was not a propitious moment to fold his hand.

Besides, Franco had his own data, and he believed in them. Nothing convinces scientists more effectively than the experience of seeing data emerge from their own experiments. In this case there were, to be sure, many questions. It turns out that neutrons are not so easy to detect, and the instruments used to detect them are sometimes tricky and undependable. In the aftermath of the Frascati announcements, experts from Italy and abroad (especially the United States) made brief visits to Scaramuzzi's lab. They gave it as their opinion that the apparent bursts of neutrons that Franco's team claimed to have detected were really artifacts caused by changes in temperature or humidity, or to some electronic problem, such as power surges in the (notoriously unstable) Frascati lab electric system. I remember during my visit that summer talking to one of Franco's young colleagues, Antonella De Ninno. "Do they think we're stupid?" she asked me angrily. "Of course we thought of all those possibilities and eliminated them!" Once the Frascati researchers were convinced they had seen the real thing, they weren't about to give up because someone had made a speech in Baltimore.

There was also a bit of wriggle room available. Pons and Fleischmann had not attended the Baltimore meeting, but Steven Jones did, and he was the first speaker. He pointed out just how small was the effect he claimed to have seen, compared to what Pons and Fleischmann were claiming to have observed. (As we have noted, the number of neutrons that had ostensibly been observed appeared to be smaller than expected—by about five orders of magnitude.) Thus it seemed possible that even if cold fusion didn't produce heat (the Pons-Fleischmann claim), maybe something *was* going on at a much lower level, producing a few neutrons (as Jones and Scaramuzzi, among others, claimed). But if cold fusion merely produced a few neutrons instead of a lot of heat, it certainly wasn't going to solve the world's energy problems. Nevertheless, it seemed at the time that there just might be two kinds of cold fusion, the "bad" kind (heat) that Koonin and Lewis had put to rest and the "good" kind (neutrons) that was still scientifically respectable. The Italian press made much of the fact that "Italian cold fusion" was of the "good" kind, ignoring the inconvenient fact that this kind of cold fusion, if it existed, would be a scientific curiosity, not an epochal discovery.

In any case, after the furor died down, cold fusion research continued in a number of places. The key to continued research is financial; to paraphrase the late California politician Jesse Unruh, money is the mother's milk of scientific research. In the United States, the government funding agencies quickly fell into line with scientific orthodoxy and ceased funding anything that smacked of cold fusion. However, the industry-supported Electric Power Research Institute decided to put up some funds, just in case. In Japan, Toyota and MITI, apparently willing to accept some short-term risk in exchange for the possibility of a big payoff later, agreed to put up a few yen. In Italy, ENEA, with its budget and prestige now resting on cold fusion, could hardly refuse to permit Scaramuzzi and his group to press on. In other places, where scientists were given modest financial support and some discretion in how to spend it, some chose to pursue cold fusion. In spite of the disapproval of the worldwide scientific establishment, some research kept right on going.

Franco Scaramuzzi's group did not devote all of its attention to cold fusion. While that work was going on, they developed the world's best device for firing frozen pellets of solid deuterium into the plasma used to create hot fusion. If hot fusion is ever to produce useful energy, this is the means by which the reactor's deuterium fuel will be replenished. They also created the sophisticated cooling device that made it possible for astronomers to carry out space-based infrared observations largely above Earth's atmospheric interference, using relatively inexpensive long-range balloon flights instead of far more costly satellites. In both of these cases, they were pursuing highly successful and demanding technical research at the very center of the scientific mainstream.

They also continued to carry out investigations into cold fusion. Reacting to criticism of the primitive technique they had initially used to detect neutrons, they purchased the best neutron-detection system in the world, essentially identical to the one used by the physicist Charles Barnes at Caltech. Going one better, they installed it in physics laboratories that had been excavated under a mountain called the Gran Sasso, a two-hour drive from Rome, in order to shield their experiment from the neutron flux produced by the cosmic rays that continually strike Earth from outer space. In the galleries under Gran Sasso, this cosmic-ray neutron background falls nearly to zero, preempting potential criticism that rogue neutrons were contaminating the experiment. Franco's team also set up an automated system to monitor the neutron counter while running the temperature of the deuterium gas cell he had developed up and down. Every week or so, a team member would drive out to the Gran Sasso lab to check out the counters, replenish the supply of liquid nitrogen, and bring back the data. No one could accuse this group any longer of being unsophisticated about neutron work. And in fact this experiment, like their own earlier work and many others blossoming around the world, produced positive results, but only sporadically. There appeared to be no dependable recipe for coaxing bursts of neutrons out of the cold fusion cell. As long as that was true, the world of respectable science was not going to pay any attention even to the "good" cold fusion.

Then they decided to pursue "bad" fusion as well. They built a well-designed electrolysis cell, capable of detecting excess heat (assuming any was produced) while obviating some of the shortcomings for which previous excess-heat experiments had been criticized. In 1992 and 1993, these experiments, too, gave positive results. The cell would produce very substantial amounts of heat (a few watts) for periods of tens of hours at a time. As in the neutron experiments, these episodes were sporadic, occurring seemingly at random, but at least they occurred only when the fluid in the cell was heavy water (containing deuterium) and never when the liquid was garden-variety water (containing ordinary hydrogen). The lack of this type of experimental control (using ordinary water instead of heavy water) had been one of the key criticisms leveled at Pons and Fleischmann. However, by this time, the scientific establishment was no longer listening.

I went to visit Franco in December 1993, after he returned from the Maui conference. While I was there, he summarized the results of the conference in a seminar presented to the physics faculty at the University of Rome (known as *La Sapienza*, the first university of Rome; there are now two more). This was in itself an unusual event. The Rome physics faculty is comparable to the physics department at a good American state university, and inviting Franco to speak about cold fusion was a daring excursion to the fringes of science for such a solid university to take. Feeling this was a rare opportunity, Franco prepared his talk with meticulous care.

At the seminar, Franco's demeanor was subdued, and his presentation was, as always, reserved and correct. Nevertheless, his message was an optimistic one for cold fusion. In essence (although Franco didn't say it in these words), each of the criticisms that Caltech's Nate Lewis had correctly leveled at the Pons and Fleischmann experiments had been successfully countered by the new experiments he described at the conference.

One of the criticisms that Nate had used with telling effect is that local hot spots often develop in electrolysis experiments (Nate is himself an electrochemist, and a consummate experimentalist). By placing their thermometer at an accidental hot spot, and by neglecting the elementary precaution of stirring the water in their cells, Pons and Fleischmann could easily have fooled themselves into thinking there was excess heat where none really existed. To counter this argument, Franco could point to the design of the cell used by his Frascati group, which carefully averaged the temperature of the entire cell rather than measuring it at a single point (many other groups had introduced mechanical stirrers into their cells).

Another objection that had been raised was that any actual heat generated in these experiments was the result of some uninteresting chemical process rather than nuclear fusion. Chemical processes that generate heat are not uncommon in electrolysis experiments. As we have seen, the strongest argument for nuclear fusion (given the near absence of the neutrons and tritium) was that the reported amount of observed heat was far too large to be due to any chemical process. That would be true, the critics replied, if cold fusion cells didn't have long dormant periods during which energy was being generated at the same time as the heat. However, all of this energy was being pumped in, and no excess heat was being produced. This suggested that the heat finally liberated in the "cold fusion" episodes might just have been chemical energy stored up during the dormant periods. In other words, far from producing more energy than was being put into them, the cells were just storing up energy and releasing it in concentrated bursts. (Not only would that be much less exciting than a discovery of controlled nuclear fusion, it also wouldn't be of much help in our struggle against the oil barons.) Now, as Franco reported, this argument could be countered as well: There were what appeared to be very careful experiments, including his own, in which the total amount of energy consumed during the dormant periods was minuscule compared to the amount of heat liberated during the active periods.

Finally, one of the most damaging criticisms of Pons and Fleischmann was that they had failed to do control experiments. Nuclear fusion (if it occurred) should have been possible (if it was possible) only when electrolysis was carried out in heavy water, made of deuterium. It should not have been possible using ordinary water, made of hydrogen. Now many groups, including Franco's, had done the necessary control experiments, and obtained the necessary confirming results (no heat in the water controls). Unfortunately, other groups reported that they did observe excess heat in experiments done with ordinary water. Franco dutifully reported these results at the Rome seminar, expressing only muted disapproval ("In my opinion, these results have not been consolidated," he said).

All of this was much less important than the fact that cold fusion experiments, if they gave positive results at all, gave them only sporadically and unpredictably. When Bednorz and Müller announced the discovery of high-temperature superconductivity in 1986, no one carped about control experiments, because once the recipe was known, any competent scientist could make a sample and test it and it would work immediately. If, at their press conference, Pons and Fleischmann had given a dependable recipe for producing excess heat, they very likely would be Nobel laureates (as Bednorz and Müller are) rather than social outcasts from the community of scientists. The essential key to returning cold fusion to scientific respectability is to find the missing ingredient that would make the recipe work every time.

Remarkably, very little has changed in the years since these events unfolded. Koonin, Barnes, and Lewis are still resolute in thinking they have cast cold fusion firmly out of the house of science, and most scientists agree with them. Nevertheless, a few scientists here and there continue to pursue it, and international meetings on the subject are held every two years or so. No one has succeeded in making cold fusion occur dependably all the time, but there continue to be enough suggestive results to keep people interested. My friend Franco has taken to doing experiments on the loading of palladium and platinum into deuterium, a respectable pursuit with only a tangential connection to cold fusion.

Recently I told this story in the philosophy course on scientific ethics that I teach at Caltech with Jim Woodward. When I finished my tale, the first question I was asked was, "Do you believe in cold fusion?" The answer is no. Certainly I believe quite firmly in the theoretical arguments that say cold fusion is impossible. On the other hand, however, I believe equally firmly in the integrity and competence of Franco Scaramuzzi and his group of coworkers at Frascati. I was concerned when I saw that Franco had gotten caught in the web of science-by-news-conference in April 1989 (although I was truly pleased that he finally got the long overdue recognition his agency ENEA owed him), and I was even more distressed when I learned that Franco and his group claimed to have observed excess heat (the "bad kind" of cold fusion) in his experiments. However, over the years, I have looked at his team's cells and at their data, and it's pretty impressive. What all these experiments really need is a thorough critical examination by accomplished rivals intent on proving them wrong. That is part of the normal functioning of science. Unfortunately, in this area, science is not functioning normally. There is nobody out there listening.

When all is said and done, the cold fusion saga offers a classic case study of how scientists, bent as they are on deepening and enlarging their understanding of nature, may convince themselves that they are in the possession of knowledge that does not in fact exist. This is not scientific misconduct, but it is a phenomenon of considerable scientific and human interest. Many mistakes were made on both sides of the great divide in the course of the cold fusion story, but they were honest errors, not examples of fraud. I suppose that if nuclear fusion really has taken place in some of these experiments, the world of conventional science will eventually be forced to take notice. If not, then the whole episode will remain a curious and instructive footnote in the history of science. Either way, it illuminates the inner dynamics of the scientific enterprise in a way that few other stories have done. For that reason alone, it will always be worth telling.

Ⅱ Six
Fraud in Physics

"The physicists have known sin," J. Robert Oppenheimer is famously said to have remarked after the atomic bombs were dropped on Hiroshima and Nagasaki. In more recent years, it had seemed that the physics community might be immune or at least highly resistant to another form of sin—that of fabricating scientific data. Nearly every case of scientific fraud in the last three decades seemed to involve biology and related sciences, not physics. In the first years of the twenty-first century, however, two high-profile cases of cheating in physics emerged into the harsh light of day. One involved the announcement and later retraction of the discovery of element 118 at Lawrence Berkeley National Laboratory. The other concerned a young researcher at Bell Labs named Jan Hendrik Schön.

Schön received his Ph.D. in 1997 from the University of Konstanz in his native Germany after spending a summer working at the renowned Bell Laboratories in Murray Hill, New Jersey. After earning his doctorate, he accepted an offer to return to the Murray Hill facility as a postdoc, although he first had to spend some time cooling his heels in Konstanz waiting for his visa to come through. Despite his youth, he had already acquired quite a reputation as a brilliant young experimental physicist, and he struck many as a prime candidate for future laurels, perhaps even a Nobel Prize.

Figure 6.1
Jan Hendrik Schön. Reprinted
with permission of Alcatel-Lucent
USA Inc.

Schön worked in the field of semiconductors, the kind of devices of which transistors are made. However, his research was in the new area of organic, or carbon-based, semiconductors, so his devices, novel and promising though they seemed, were as yet at the strictly experimental stage. Organic semiconductors are generally lighter, more flexible, and less expensive than the inorganic kind—all desirable advantages—but they also have more resistance and therefore conduct electricity less well. But miraculously, first at Konstanz and then at Bell Labs, Schön seemed to leap over every technical hurdle in his path. He reported, for example, using field-effect doping—applying very large electric fields to change the electron concentration in his samples—to induce such remarkable phenomena as superconductivity and the Quantum Hall effect, the first time such results had ever been reported for organic semiconductors. Other researchers, despite repeated attempts, had not been able to generate high enough fields to detect these miraculous effects because of electrical breakdowns in the insulating layers that are essential for such experiments. But Schön, using a humble apparatus in Konstanz, had allegedly managed to produce insulating films of aluminum oxide that conferred unprecedented resistance to breakdown.

Between 1998 and the summer of 2001, Schön, working with a total of twenty collaborators, produced on average one research paper every eight days. He was soon hired onto the permanent staff of Bell Labs. Later he was offered the directorship of a Max Planck institute in Germany, which he was considering when he submitted a new paper to *Science* in December 2001.[1]

It was at this point that the wheels started to come off. In his *Science* article, entitled "Field-Effect Modulation of the Conductance of Single Molecules," Schön and two coauthors announced that they had successfully fabricated a single-molecule transistor—the logical endpoint of Moore's law, which holds that the number of transistors that can fit on a single chip should double every eighteen months or so. It was this seemingly triumphant breakthrough that touched off an unsuspension of disbelief. Among the first to raise the red flag was Cornell University physics professor Paul McEuen, whose suspicions were aroused by data that struck him as too perfect, including accounts of different experiments that had identical noise—the usually random background disturbances that show up in any experiment. Soon other scientists, researchers at Bell Labs among them, were calling attention to other anomalies. Matters came to a head in the spring of 2002, when Bell Labs appointed a committee, chaired by Stanford physics professor Malcolm Beasley, to look into these allegations and, in a rare instance of openness in the murky field of scientific misconduct, made it clear from the outset that it intended to publicly release the results of its investigation.

The committee's report was duly released on September 25.[2] It detailed twenty-four specific allegations and found that Schön had committed scientific misconduct in at least sixteen of them. The report further concluded that Schön, his twenty collaborators notwithstanding, had carried out all of his experiments ("with minor exceptions") alone and that "proper laboratory records were not systematically maintained." Virtually all of his raw data files had been erased from his computer, and all of his original samples had been either destroyed or discarded. With only the slightest of misgivings, the report exonerated all of his collaborators, saying that the scientific community had not yet come to grips with the issue of coworker culpability in such cases. Schön himself was immediately fired by Bell Labs and has since dropped out of sight. The University of Konstanz revoked his doctorate in 2004.

The Schön case raises a number of issues. To begin with, it is amazing that when suspicions concerning his research first surfaced, Bell Labs had no formal policy on how to handle cases of research misconduct. By this time, all American universities that accepted federal research funds were required to have such policies, but Bell did not receive federal funding. The attitude at the lab seems to have been one that was common in the universities a couple of decades earlier—such egregious behavior couldn't possibly happen here, so why do we need such a policy?

In the absence of any official lab protocols, the Beasley committee took the prudent course of following the guidelines on scientific misconduct that the federal government had set down for the nation's universities. Under these guidelines, proof of scientific fraud did not need to be established beyond a reasonable doubt, as in a criminal case, but rather by a preponderance of the evidence. A more difficult issue related to the responsibility of the other authors on Schön's plentitude of papers. Ultimately, the Beasley report defined this as a matter not of scientific misconduct but of "professional responsibility," and concluded that "no clear, widely accepted standards of behavior exist," because it is an issue that "the scientific community has not considered carefully."

In fact the issue here is trust among scientists. Collaborations take place precisely because different scientists bring different skills and perspectives to the table. If they must also be responsible for obsessively looking over the shoulders of their fellow researchers, collaborations will fall apart, and much damage will be done to science. Although it makes one uneasy that none of Schön's many collaborators ever thought to question the validity of his work, scientific fraud is so rare that probably none of them suspected any wrongdoing until the very end.

In chapter 1 I outlined three motivating factors that seem invariably to be present in cases of scientific fraud, noting that these have been extrapolated from the largest pool of misconduct cases: namely, those occurring in biomedicine. How well do these factors hold up in the context of fraud in the physical sciences? In this case, they seem to do pretty well. Was Schön under career pressure? Absolutely, as is everyone at a place like Bell Labs, perhaps made all the more brutal by the intensely competitive nature of the field he was in.

Did he believe he knew the right answer? That is, did he believe that all the remarkable effects that he claimed to have found really existed? Apparently he did. In a response appended to the Beasley report, Schön admitted having made mistakes but maintained that "I have observed experimentally the various physical effects reported. . . such as the Quantum Hall effect [and] superconductivity in various materials. . . . I believe that these results will be reproduced in the future." (For the record, none of Schön's results has been reproduced.)

Finally, was Schön working in a field in which results are not easily reproduced? He was. Results in the field of semiconductor devices are notoriously sample-specific, depending crucially on the skill and luck of the person who prepares the sample. Failure to reproduce any given result in any given sample is not considered proof of anything. Nobody could prove Schön had cheated just by attempting to replicate a particular result and demonstrating that it didn't show up in a particular sample. In fact, until McEuen and others started noticing duplications in the published data, no one complained about the results. So my theory survives to be disproved another day.

Now we come to the case of Victor Ninov, who in 1996 arrived at Lawrence Berkeley National Laboratory (LBNL), a federal facility run by the University of California. In its heyday, the Berkeley facility, using the lab's 88-inch cyclotron, or particle accelerator, had synthesized a series of elements lying beyond uranium, the last naturally occurring element—each discovery a cause for celebrations and congratulations. As the new century approached, however, that charmed era seemed to have ended. No new elements had been unveiled in some years using the cyclotron, and it was generally acknowledged that further breakthroughs would require new technologies.

At LBNL Ninov, who had established a good reputation in heavy-element research during several years at the German Heavy Ion Research Laboratory in Darmstadt, oversaw the construction of the Berkeley Gas-Filled Separator (BGS), an instrument designed to sort through the debris of nuclear collisions in the cyclotron and potentially discover new heavy elements. In 1999, shortly after its completion, just such a prospect appeared

Figure 6.2
Victor Ninov, Lawrence Berkeley National Laboratory, 2000. Courtesy of Lawrence Berkeley National Laboratory—Roy Kaltschmidt, photographer.

in the writings of Robert Smolańczuk, a nuclear theorist visiting the lab from Warsaw. He proposed that element 118—an unstable behemoth with 118 protons in its nucleus—could be created by scattering krypton 86 from lead 208 (technical parlance for, essentially, firing a beam of krypton atoms at a lead target). The predicted probability of this reaction's yielding element 118, assuming that the BGS and the cyclotron were operating at their standard rate, was about one event per week. The new element, being unstable, would almost immediately start to decay by giving off alpha particles, and its fleeting presence would be detected by observing these signature decays.

The BGS group undertook to do the experiment, and in the August 9, 1999, issue of *Physical Review Letters,* it reported having observed three decay chains—that is, a trio of nuclear reactions decaying in the manner expected of element 118.[3] For the discovery of a new element to be firmly established, the results must be reproduced by another group. Three groups, in

Germany, France, and Japan, soon undertook to do so. All these research teams were working with a more powerful apparatus than the 88-inch cyclotron and should have been in a position to observe a greater number of 118 decay chains. When none of the groups succeeded in producing a single positive result, suspicions were raised.

The Lawrence Berkeley group swung back into action and produced one more positive result. But by this time the physics community was skeptical, and in 2001 an investigation, chaired by the LBNL's Gerald R. Lynch, got under way. Ultimately, the panel found clear evidence that the data from all four Berkeley events had been fabricated and concluded that the only person in a position to fake them was Victor Ninov, who had assiduously shepherded the experiment from beginning to end.

The definitive judgment against Ninov came in the form of an internal March 2002 report commissioned by LBNL's head, Charles Shank.[4] This document, some 217 pages long, concluded that Ninov had indeed been guilty of scientific misconduct. The document was confidential (and still is), but I, along with a number of journalists, was able to obtain a copy of it under the California Public Records Act. Imagine my surprise when I learned from the title page that the committee had been chaired by my Caltech colleague, professor of physics Robbie (Rochus) Vogt. I immediately asked Robbie to present the case to the Scientific Ethics class that Woodward and I were co-teaching, and he said he would be happy to do so. He joined the class a week or two later and, after first providing a detailed technical account of how the BGS worked, launched with considerable relish into the nitty-gritty of his committee's report. At the heart of that report was the following summary:

Figure 6.3
Caltech physicist Rochus Vogt,
1983. Courtesy of Robert Paz/
Caltech Public Relations.

There is clear evidence to conclude that Dr. Ninov has engaged in misconduct in scientific research by carrying out this fabrication. He was the only collaboration member doing analysis in 1999, and he was the one who announced both the 1999 chains and the initially claimed decay chain in 2001. If anyone else had done the fabrication, Dr. Ninov would almost surely have detected it.

The report went on to say that "Dr. Ninov, although proclaiming his innocence, did not provide any substantive basis for changing the committee's conclusions as to his role in the fabrication of data." Ninov, who had been placed on indefinite paid leave in November 2001, pending the investigation's final outcome, was dismissed from his post in May 2002, while continuing to vigorously deny any wrongdoing. Even though full accounts of the affair appeared in *Physics Today*,[5] *Science*,[6] and other publications, he found a job as an adjunct professor of physics at the University of the Pacific, which apparently was unaware of his recent history. He is no longer listed on the faculty of that institution.

Do the circumstances of the Ninov case violate our earlier assertion that scientific misconduct occurs in fields where precise reproducibility is not expected? This episode of fraud unfolded in a field where precise reproducibility is demanded— and that simple fact was its undoing. Because Smolańczuk's theory predicted it, Ninov no doubt expected element 118 to exist. Thus even if he fabricated data to make his "discovery," he expected that his findings would be confirmed and he would get credit for it. Indeed, element 118 may yet be found to exist, but Smolańczuk's theory of the probability of producing it by scattering krypton 86 from lead 208 has turned out to be inaccurate. In effect, Ninov took the reproducibility criterion and stood it on its head. Unfortunately for him, it didn't stay in that uncomfortable position for very long.

The Schön and Ninov cases put scientific misconduct back on the front pages of the newspapers, and this time it was physics that was on the firing line. Inevitably, there was much debate and soul-searching in the physics community about how common the phenomenon might be and what to do about it. That debate is still going on.

⌐ Seven
The Breakthrough That *Wasn't*
Too Good to Be True

This chapter has its origins in a talk on the discovery of high-temperature superconductivity that I gave in 1989 as part of the Caltech Watson Lecture series, which regularly presents public lectures by Caltech professors. The subject captured my attention not only because low-temperature physics (to which superconductivity belongs) is my own research field but also because it offers a rare instance in which a physical phenomenon that just about everyone had always known to be impossible turned out to be not only possible but quite commonplace once a couple of persistent and enterprising researchers had ferreted out its existence. In an ideal world, no scientist could ever fail to be charmed by the realization that once again, nature's ingenuity had trumped ours. In the real world, of course, matters are a little more complicated, and the discovery was initially greeted with a certain amount of disbelief (by those who had "always known," etc.) and consternation (by those who knew that whatever else might prove to be the case, they had not been the ones to make the discovery). But once the initial hype was over and the definitive confirmation came in, the scientific community settled down into its praiseworthy pattern of following up and amplifying on a genuinely thrilling eureka moment (my own lab was one of many that successfully replicated a number of the initial experiments). With the benefit of hindsight, we can also see the ways in which this story both resembles and departs from the cold fusion controversy that gripped the scientific community a few years later. Taken together, the two offer classic and contrasting case studies into how seemingly incredible discoveries are either assimilated into the body of scientific knowledge or ultimately rejected by it as untenable. Twenty years have passed since I first presented and subsequently published this talk. What follows is the original piece; I then append an epilogue to bring various matters up to date.

⌐F The two researchers who discovered high-temperature superconductivity in 1986 were working in a field in which most people had given up hope. In fact, J. Georg Bednorz and Karl Alexander Müller at the IBM Zurich Research Laboratory, in Rüschlikon, Switzerland, had to disguise their work from their own supervisor in order to be able to do it. And even after they had made their discovery, the poor guys had to wait nearly an entire year before they got their Nobel Prize. The discovery[1] was quickly confirmed in unexpected places such as China and Japan, and especially in Houston, Texas, and Huntsville, Alabama, where Professor Paul C. W. Chu and his students and associates and former students not only confirmed the discovery but soon found a new class of materials that became superconducting at even higher temperatures. And that's when all the excitement really began.

The program for the March 1987 meeting of the American Physical Society in New York City had gone to press in December 1986, before the discovery was announced, so it contained nothing about high-temperature superconductivity. But the organizers of the meeting, sensing that there was some interest

Figure 7.1
Karl Alexander Müller
(left) and J. Georg
Bednorz, 1987.
Courtesy of IBM
Corporate Archives.

in the subject, obligingly arranged for a special evening session to be held in case anyone had anything to present about the topic. Nearly 4,000 people attended the session, which began at seven in the evening and finally broke up at six the next morning. The *New York Times* ran a front-page story calling it a "Woodstock for physicists."[2] One month later, the whole scene was repeated at the annual meeting of the European Physical Society in Pisa, and although I was in Italy at the time, I declined to go. Once had been enough.

My favorite remark to come out of all this was made at a federal conference in Washington, where Clayton Yeutter, the U.S. trade representative, said, "Chief executive officers are paid to succeed." In other words, our business and industry leaders were supposed to be on top of this situation. This was in reference to superconductivity, a subject that had been a mere scientific curiosity a few months before. Suddenly this curiosity had become cardiac country for CEOs, and the new question was: What's going on here? To try to answer this question, I'll first explain what superconductivity is; then I'll define what we mean by high temperature in regard to superconductivity; then I'll describe what the new discovery is; and finally I'll discuss whether anything might be made of it.

The phenomenon of superconductivity was discovered in 1911 by the Dutch physicist Heike Kamerlingh-Onnes. Three years earlier, Kamerlingh-Onnes had succeeded in becoming the first person ever to liquefy the element helium, and in liquid helium he had a bath in which he could cool things to the lowest temperature ever achieved on Earth. So any measurement he decided to make would be the first measurement ever made at this new lowest temperature. The measurement he decided on was the resistance of a sample of metal. He chose mercury, because he could distill it and make it very pure. The experiment is done by passing an electric current through the sample and measuring the voltage that develops across the sample in order to push the electric current through it. The ratio of voltage to current is called the resistance. Kamerlingh-Onnes wanted to find out how the resistance of mercury behaved as a function of temperature when it was cooled down to this very low point.

It was well known at the time that at higher temperatures, as the temperature went down the resistance went down in a nice, smooth curve. You might wonder why this man, with a whole new world to explore, would choose to add a few more points to a well-known curve. The answer, I think, was that he didn't expect to do that at all; he expected to make a spectacular discovery. His mental image of how a metal worked would

Figure 7.2
Heike Kamerlingh-Onnes,
circa 1911. Courtesy of
Museum Boerhaave.

have been much the same as ours today—a metal is basically a container full of a fluid of free electrons that can move around inside and give the metal its familiar properties, such as its shiny surface, electrical conduction, and so on.

Kamerlingh-Onnes also would have known that all fluids freeze if you cool them down enough, so he might have thought that if he got a metal like mercury cold enough, the electron fluid would freeze and it would cease to be a metal. If that happened, then at the point where it ceased to be a metal, the resistance would suddenly jump up to infinity. I think that must have been the dramatic discovery he expected to make. But when he made the measurement, what he found instead was exactly the opposite—the resistance jumped down to zero. And that was the discovery of superconductivity.[3]

Superconductivity has, we now know, three principal properties associated with it. One is that the electrical resistance is zero; the second is that superconductors are destroyed by magnetic fields; and the third is called tunneling.

In saying that the electrical resistance is zero, I mean really zero, not just very small. The way to find that out is not by Kamerlingh-Onnes' method; in that case, all you find out is that the resistance is smaller than you're able to measure with your instruments. The better way to do it is to take a doughnut of superconducting material, such as lead, and induce a current to run around in the ring with nothing driving it. If you do this with the very purest and best ordinary conductor—for example, very pure copper at low temperature—the natural resistance of the material would cause the current to decay away to zero in a tiny fraction of a second. But when this experiment is done with superconducting lead, the current flows with no noticeable decrease for a period of years. So the resistance to the flow is really zero.

Superconductors are destroyed by magnetic fields. If you apply a magnetic field to a superconductor, the superconductor actually expels the magnetic field—keeps it outside so that the inside has no magnetic field and can remain superconducting. Of course, if you make the magnetic field stronger, it becomes harder for the superconductor to expel it, and if you make it strong enough, the superconductor can no longer do it. The field collapses into the material and the material ceases to be superconducting.

In the early 1960s, however, somebody discovered that certain superconductors (not all of them, but some) actually could support extremely large magnetic fields before they were destroyed. Nobody remembers who discovered high-field superconductors (well, I'm sure the discoverer remembers, but I don't), but I think in the long view of history that might turn out to be a more important discovery than that of high-temperature superconductivity.

If you have a superconductor that can exist in a high field, it can also be used to create a high field. To do that, you make the superconductor into a wire and make the wire into a coil; then you run a very large current through the coil, creating a magnetic field inside the coil. So the second property is that superconductors can be used to create magnetic fields, and in some cases very large magnetic fields.

The third property of superconductivity is known as tunneling. If you put two pieces of superconductor very close together, then it's possible for some of that supercurrent (that is, the current than can flow with no resistance) to leak across from one to the other. Then, if you repeat Kamerlingh-Onnes' measurement, you find the same result. The current flows and there's no voltage at all across the circuit, across what we call the tunnel junction. Unlike the case of a single piece of superconductor, however, when the current is flowing through a tunnel junction, the relation between current and voltage becomes exquisitely sensitive to tiny outside influences, such as very small magnetic fields, electromagnetic radiation, and so on.

Now that we know what superconductivity is, let's go on to temperature. In Kamerlingh-Onnes' mercury sample, the superconducting transition occurred at a temperature of 4 kelvins (K). Absolute zero—that is, the lowest temperature that has any meaning, the temperature of a body from which all possible energy has been extracted—is 0 K. That's very convenient; it means we never have to deal with negative numbers. (In the familiar Fahrenheit scale, absolute zero is −459°.) The size of a degree in the Kelvin scale was fixed by the very sensible idea that the freezing point of water should be 273 K. Once you've made that decision, then it follows that normal room temperature is about 294 K and the temperature at which air liquefies is about 80 K. At this temperature, which is −330°F, most of the air in the room would form a puddle on the floor. Of course, if the temperature in the room were −330°F, other things would happen that don't bear thinking about.

In seventy-five years of work on superconductivity, scientists managed to push the maximum temperature at which it occurred from 4 K up to 24 K. It held at that point for a long time, which is why Bednorz and Müller were working in an obsolete field. The breakthrough that earned them the Nobel Prize was the discovery of material that became superconducting at 30 K. The big discovery by Chu and his associates that caused all the excitement was a material that became superconducting at 93 K, the crucial point here being that it's above the temperature of liquid air.[4] So now you could cool a sample in liquid air and get a superconductor, whereas before you had to cool a sample in liquid helium.

The cost of the required refrigerants is a good illustration of the difference between these temperatures. For the old-fashioned kinds of superconducting materials, you have to use liquid helium. Liquid helium delivered to our door at Caltech with the morning milk costs ten dollars per liter, just about the cost of cheap vodka. It even looks a lot like cheap vodka—a colorless fluid. We get liquid nitrogen (liquid air and liquid nitrogen are pretty much the same thing, since air is 80 percent nitrogen) delivered to the door for about eleven cents per liter. That's a lot less than what we pay for bottled drinking water in the lab. Now we know that a number of high-temperature materials will be superconducting at the temperature (a little below 80 K) of liquid nitrogen.

The first superconducting material, as I mentioned earlier, was mercury. But it turned out that many of the metallic elements (lead, tin, indium, and aluminum, for example) become superconducting at low temperatures. However, the best ordinary conductors, such as copper, silver, and gold, never become superconducting no matter how low the temperature gets. Neither do the magnetic materials—iron, cobalt, and nickel. (Magnetism is inimical to superconductivity.) High-field superconductivity was discovered in the early 1960s in the material niobium-tin; and other high-field superconductors, including niobium-titanium, were discovered later.

But the new high-temperature materials are quite different from the metals and alloys that become superconducting at lower temperatures. Certainly they have more alien-sounding names. Bednorz and Müller made their breakthrough discovery (30 K) while working with lanthanum copper oxide that contained a small amount of strontium impurity. Chu and his colleagues made their 93 K discovery with the material yttrium-barium copper oxide, and this was followed by a couple of discoveries of newer materials at even slightly higher temperatures—bismuth-strontium-calcium copper oxide at 107 K and thallium-barium-calcium copper oxide at 125 K.

I remember that shortly after the initial discoveries, and in the midst of all the excitement and speculation, my physics colleague Richard Feynman dropped by my office one day. We discussed the new findings a little bit, and he predicted that the highest-temperature superconductor would be based on the element scandium. His first argument was that scandium came from roughly the right part of the periodic table to replace the exotic materials in these compounds, and his other argument was that of all the elements in the periodic table, scandium is the only one for which no purpose had ever been found. Naturally, I shared this observation with my colleague George Rossman, who is a Caltech professor of mineralogy with ninety-two personal friends on the periodic table (and that only counts the stable ones), and he said that that was completely wrong—there was some use for scandium . . . but he couldn't remember what it was. The usefulness of scandium aside, these new superconductors are complicated composite materials of a type technically called ceramics. What we're dealing with here is pottery.

Within a month of the announcement of Bednorz and Müller's discovery, my research group had made our own measurement of the superconductivity of lanthanum-strontium copper oxide in our lab. That's not a tribute to how clever or fast we are but to how easy the stuff is to make. It was made in a hundred laboratories around the world as soon as the discovery was announced. We also measured the resistance of the yttrium-barium copper oxide material, which drops to zero at about 90 K.

ⅠF So that's what high-temperature superconductivity is all about. Why has it caused such excitement? I think there are a number of reasons for it. The first one is the history of the subject. For seventy-five years, people tried mightily to find materials that would become superconducting at a slightly higher temperature, but after inching it up to 24 K, nobody could move it up at all for the next fifteen years. Just about everybody gave up on the possibility of getting any higher. There was even a theory that predicted that the highest possible temperature for superconductivity was 35 K. Then all of a sudden came the Bednorz and Müller discovery of 30 K, followed by Chu's discovery at 93 K, and everyone began to expect to see great things at up to room temperature within a month at the most. So this generated a great deal of excitement.

The second reason has to do with a mystery. We physicists are taught to believe that there is no such thing as a frictionless surface or a liquid with no viscosity or, for that matter, a conductor with no resistance. Kamerlingh-Onnes' discovery of superconductivity shattered part of that received wisdom—sure enough, there is a conductor with no resistance—but because this phenomenon existed only under conditions of extreme low temperature, the thinking was that perhaps that compensated in some sense for its bizarre behavior. Now, more than fifty years later, we've suddenly got stuff becoming superconducting at 100 K and above. Somehow it seems wrong; it goes against all our instincts, and furthermore we don't understand why it occurs. There are lots of theories—in fact, one theory for every theorist. What is lacking is a consensus; there is no agreement among physicists as to what this is all about.

Finally, there is a good deal of excitement because some people suspect that somebody may someday find something to do with this stuff. There may be some practical application. If there is to be any practical application of high-temperature superconductivity, it will make use of superconductors' three properties: zero resistance, the ability to create magnetic fields, and tunneling.

When you think of something that conducts electricity with no resistance at all, the first idea that comes to mind is electrical power transmission. Suppose we could make the national power grid out of superconducting material. Wouldn't that be wonderful? In order to think about this seriously, we have to compare three possible types of systems. One is the system we have now; the second is a national power grid of the old-fashioned, low-temperature superconductor; and the third is a national power grid of a high-temperature superconductor.

The first thing to think about is what we refer to as "losses" in the system. The electric power lines have normal resistance, and passing an electric current through a resistance causes heating, which amounts to a loss of energy. In our present power grid, these line losses amount to about 10 percent of all the power generated in the United States. If you could substitute a power grid made entirely of superconductor, one that was completely free of resistance, you would retain that 10 percent.

In fact, as early as the 1960s, extensive engineering studies were done to examine the possibility of a superconducting power transmission system. This was long before the discovery of high-temperature superconductors, and the plans in those days called for cooling the system with liquid helium. One of the findings of those studies was that the cost of refrigeration was relatively small: You don't have to cool the system down from room temperature every day; if you get it cold once, it will stay cold forever. You just have to compensate for the small amount of heat that leaks into a well-insulated system by building refrigerators that tap into a little bit of the power that the system is transmitting. Thus the additional savings we would achieve by using high-temperature superconductors represents only a fraction of the already negligible cost of refrigeration.

On the other hand, the superconducting materials themselves are expensive, and notoriously difficult to make into the kinds of wires needed for a superconducting power transmission system. The problem—and it was a major problem—was materials fabrication, and what we've now got is material that is even more balky and unwieldy to deal with than the original stuff. All things considered, what we have here doesn't seem to be a big step forward for high-power transmission.

The second of the properties that we might want to exploit is the ability to make high magnetic fields. The unit that most scientists use to measure magnetic fields is the gauss. (Actually, we're not supposed to use gauss anymore; we're supposed to use a tesla, which is 10,000 gauss, but most of us are old-fashioned and still use gauss.) The Earth's field has a magnitude of about 0.5 gauss. The saturation field of an iron-core electromagnet is 20,000 gauss, and this was the strongest field you could get in the laboratory back in the sixties. There was perhaps one laboratory in the world that, by dedicating an enormous facility to that purpose, was able to create a field as high as 100,000 gauss for a scientific experiment. The great discovery in the early sixties of high-field superconductors was that some of these can actually be used to create a field of 100,000 gauss, with the consequence that today you can just walk into your friendly neighborhood superconducting magnet store and buy a magnet off the shelf, and you've got 100,000 gauss in your laboratory.

The new superconductors are also of the high-field type. Furthermore, since the strength of the field is directly related to the temperature, the higher the temperature, the higher the field. Nobody knows yet how large a field will be possible with the new superconductors, but it's quite possible that it will be as big as half a million or even a million gauss. So when I say that we may have high fields available sometime in the future, I mean very, very high fields.

So what might we do with very large magnetic fields? My favorite idea by far is the magnetically levitated superconductor train. The track on which this train rides is just an ordinary material—steel or some other ordinary metal, not a superconductor. But inside each car of the train there are powerful superconducting magnets. These magnets get turned on one after the other, like the lights on a movie marquee. This fools the track into thinking that magnets are running backwards along the track. The track doesn't like that; it resists the motion of the magnets, and that gives the train an impulse to move forward. The track also repels the moving magnets, and so the whole train lifts up off the track by a few centimeters and rockets forward at a speed of 500 kilometers per hour. (That's an inch off the track at 300 miles per hour if you live in the United States.)

If you've ever been in a truly high-speed train, such as the TGV in France or the Tokyo-Osaka bullet train in Japan, then you know how much fun it is to ride along in a train at 130 miles per hour. This thing will go just about twice that fast. It will put Disneyland out of business. If they ever build it, I will certainly be the first in line to buy a ticket, but I don't think we should start getting in line quite yet. There's no problem with the technology; the Japanese have had a few-kilometer, test-bed levitated-train track running for many years to take VIPs on little rides to show them that it works. But there are a number of problems with building it, principally the expense. Among other things, the track must be very straight and very well maintained. When you're traveling at 300 mph one inch above the track, you don't want any bumps and you don't want any hairpin curves. The capital investment for buying the right-of-way and building and maintaining the track and so on would be enormous, and because we already have ways of getting from one place to another—cars, airplanes, and so on—there's no driving economic force to build this thing. Also, there's some concern about having 100,000-gauss magnets on a public conveyance. There is no conclusive evidence that large magnetic fields pose a health risk, but the jury is still out and a number of epidemiological studies are under way. In any case, it is quite likely that concerns over this issue would need to be factored into any plans for constructing this type of train.

If high-temperature superconductivity is going to have applications, one of the likeliest places is in space. For one thing, economics in space is different from economics on Earth. If a communication satellite, for instance, costs $100 million to build and you can save a little weight by using a component that costs $1,000 to replace a heavier component that costs ten dollars, odds are that you will do it. Weight is everything and cost means almost nothing. Also, there's a possibility that in a well-designed spacecraft, the ambient temperature will be low enough so that these materials will be superconducting without any active refrigeration at all.

Another problem associated with magnetic fields is a phenomenon you're familiar with if you've ever tried to open the switch on a circuit containing a big electromagnet. What happens is that the switch sparks over, because the magnetic field doesn't like to be turned off quickly. If you do that with a very big magnetic field, the spark can be so violent that it vaporizes the switch itself. To find out what kinds of problems that could cause, I called up a friend of mine who is a knowledgeable and experienced power engineer and asked for his opinion. I can't put in print exactly what he said, because these engineers can be pretty salty, but the gist of it was that "It's likely to be a challenging task."

The third property of high-temperature superconductors is tunneling. If I put two pieces of superconductor very close together, some superconducting current can leak from one to the other, but the amount is acutely sensitive to such things in the environment as magnetic fields (including very weak ones) and electromagnetic radiation. That means that you can use this tunneling property as a detector, and in fact conventional superconductivity is regularly used to detect infrared radiation. For example, orbiting satellites use superconducting tunnel junction detectors to identify infrared radiation, either in devices that look upward toward the sky for purposes of astronomy or downward toward the Earth for purposes that are usually military secrets.

The fact that the conducting state of one of these junctions is sensitive to small magnetic fields means that you can use small magnetic fields to manipulate that state—to turn it on or off. This makes it possible to use these junctions for electronic logic and memory devices. In other words, it's possible to build a superconducting computer instead of the kinds we have now, which are based on semiconductors.

A number of companies, including IBM and AT&T, were actively involved at one time in superconducting-computer research projects, but they had all discontinued them by 1983. This wasn't because superconducting technology was not advancing; it was. The reason was that the semiconducting technology was advancing much faster, and it seemed highly unlikely that superconductivity, with all its material-fabrication problems, would ever catch up. And as I mentioned earlier, the fabrication problem with these new superconductors is even worse than with the old ones. In the long run, however, the superconducting computer will ultimately be the computer of choice. The reason is simple: The whole game in computers is to make them fast, and to do that, you have to make them small, because it takes time for the signal to travel any distance. So you want to pack the elements of the computer as tightly together as you can. When you pack a lot of semiconducting elements together, as is done in present computers, each of the elements is generating heat, because it's an electrical device, and devices using electrical conductors generate heat. If too many of them are packed too closely together, the computer melts and that's no good. That's the ultimate physical limitation on how small, and therefore how fast, a computer can be. That limitation does not apply to superconducting circuit elements, because they don't dissipate heat the way semiconductor circuit elements do. So in the long run the superconducting computer will win—but it might take a very long time.

⛶ Let me end with a brief summary of human history. It starts with the Stone Age, when pottery was invented. It goes on to the Bronze Age, the Iron Age, modern times, and finally to the discovery of high-temperature superconductivity—which is the second coming of the age of pottery.

This, of course, reminds me of a little story about the Second Coming and the Pope. I hope no one finds this offensive; I've spent quite a bit of time in Italy and consider the Pope to be sort of a friendly neighbor. The story takes place in the Pope's Vatican chambers, where he's talking with two of his assistants, who are both monsignors. One of the assistants notices a strange light coming from the window, so he goes to the window to investigate, and he sees the Second Coming. But he doesn't believe it at first, because he's trained to be skeptical—not to accept things on first sight but to think about things and analyze them. So he does all those things and finally decides there's no doubt about it—what he's seeing is the Second Coming.

So he calls his colleague over to the window, and his colleague looks out and says, "You're absolutely right. There's no doubt about it. This is the Second Coming." So the two of them spin around, fall to their knees, and say, "Your Holiness, it's the Second Coming."

The Pope races to his desk, sits down, and starts typing madly on his typewriter. One of the monsignors asks, "Your Holiness, what are you doing?" And the Pope says, "Well, I don't know about you, but I want to look busy."

I don't know whether high-temperature superconductivity is really the second coming. But I do know that a lot of people are looking busy.

Epilogue

What has happened in the years since I gave that 1989 lecture is really not all that exciting. The highest temperature superconductor functions only at about 135 K, or 165 K for the same complicated material under pressure, which is still far below ambient temperatures anywhere on Earth.[5] The highest current that can be passed through a high-temperature superconducting wire has increased, but only for short lengths of wire, not for long ones that can be wound into coils. However, the tunneling property I spoke about twenty years ago has been exploited, and one can now find high-temperature superconducting tunnel junctions in a number of applications, particularly scientific ones. The very high fields that seemed to be in the offing have largely not materialized, because various other things break down before the superconductivity itself does.

We can also identify a handful of other changes. In the last twenty years, the cost of liquid nitrogen has risen from roughly eleven cents to eighty cents per liter (the price of liquid helium has actually declined a bit). The speeds of the Tokyo express and TGV bullet trains have likewise appreciated, to about 170 mph and 200 mph, respectively. And my comfortable assumption that there would be little or no economic incentive to construct superfast trains in a world where cars and airplanes reigned supreme has suffered the fate of many such confident pronouncements, along with our faith in the inviolability of cheap fuel prices. In fact I write this amid talk that policy makers are taking a fresh look at the prospect of building such train systems in the United States, including one that would connect my region, the L.A. basin, with—where else?—Las Vegas. Still, I have no hesitation in rashly venturing once again into unknown territory with the prediction that it will be quite some time before high-temperature superconductivity produces any sort of revolution in the way we live. In the meantime, research into potential applications continues, and the phenomenon's underlying dynamics pose a superb scientific problem, crying out for an explanation. Whether a dazzling explanation or a dazzling application will come first remains an open question, and one that is likely to tantalize us for some years to come.

⬚ Eight
What Have We Learned?

We have been through quite a lot, in this little book, having to do with honorable and dishonorable conduct in science. Our first chapter introduced our general framework, elaborating on various theories of how science works, particularly those of Bacon and Popper. To this day some people teach the scientific method as if Bacon's dictum were the final word. But it isn't. We sketched out three conditions that are generally present when scientific fraud occurs, and considered fifteen plausible sounding ethical principles, all of which would be damaging to science if anyone actually tried to apply them. We then revealed in all their glory those hidden twin pillars of the scientific enterprise—the Reward System and the Authority Structure.

Next we considered the somewhat controversial case of Robert Millikan. In recent times he has been accused of manipulating his data to give the answer he wanted. However, a careful analysis of his procedure tells a somewhat different story. He manipulated his data all right, but the intent was not to deceive, but to arrive at the most accurate value possible for the charge on an electron. If his experimental methodologies, using far more primitive instruments than we currently have at our disposal, strike us today as somewhat unorthodox, his results were anything but heretical. The final value arrived at by Millikan is within his cited 0.2 percent error of the modern value.

Having salvaged the reputation of Caltech's patron saint, we turned to two related cases in which fraud actually occurred at Caltech—those of Vipin Kumar and James Urban, both of whom worked in the laboratory of biology professor Lee Hood. Both cases were investigated under the Institute's then quite new rules on research misconduct, and both scientists were found to have committed misconduct. And yet, the ultimate outcomes of these cases were quite different: Urban, who had left Caltech for a faculty job at the University of Chicago, was forced to resign his position there and has since dropped out of sight. Kumar, however, found defenders within the scientific community who accepted his view that he was to some extent a victim of inadequate guidance and supervision in an exceptionally busy laboratory. He has since returned to an active life in science.

We have also seen how the United States government got into the act under pressure from Congress, appointing a veritable alphabet soup of anti-fraud SWAT agencies, commencing with the Orwellian-sounding Office of Scientific Integrity (OSI). The OSI later morphed into the Office of Research Integrity (ORI), both offices reporting ultimately to a succession of overseers, including Health and Human Services. Eventually, in a blaze of publicity, the OSI/ORI undertook to investigate cases involving colleagues of David Baltimore and Robert Gallo. They professed in both instances to have ferreted out serious misconduct, but both decisions were embarrassingly overturned by appeals boards, leaving everyone exonerated except the ORI itself. Ultimately, however, the federal government, acting in tandem with seasoned scientists from a variety of disciplines and institutions, produced a set of regulations that have served as useful and realistic guidelines for dealing with cases of alleged scientific misconduct.

Then into this seemingly black and white arena of Good Science versus Fraudulent Science comes the strange and complex case of cold fusion. Can it be that nuclear fusion takes place on a tabletop when deuterium is subjected to electrolysis using electrodes of platinum and palladium? Absolutely not! was the verdict of very highly regarded physicists and electrochemists, who debunked the phenomenon on both experimental and theoretical grounds, and their view prevails today within the scientific community. But a handful of scientists persist even now in believing that cold fusion can and does occur and continues trying to prove it. Every time I've spoken on the subject of scientific fraud, I've been asked whether cold fusion was an example. My answer is always the same: It wasn't. It may turn out to have been a wrong-headed notion, as most mainstream scientists believe today, but mistaken interpretations of how nature behaves do not and never will constitute scientific misconduct. They certainly tell us something about the ways in which scientists may fall victim to self-delusion, misperceptions, unrealistic expectations, and flawed experimentation, to name but a few shortcomings. But these are examples of all-too-human foibles, not instances of scientific fraud.

My three conditions that always seemed to be present when scientific fraud occurred appeared to be threatened in 2002 when two new cases turned up in the pristine field of physics. The three conditions are that the scientist feels himself to be under career pressure, he thinks he knows how the experiment would turn out if it were done properly, and he's working in a field where precise reproducibility is not expected. This last condition seemed to favor biology as the arena where fraud could take place. In the case of Jan Hendrik Schön, the three conditions fit like a glove The case of Victor Ninov, however, was quite different. Ninov had apparently discovered a new heavy element with atomic number 118. The first two conditions fit him well enough, but the third was turned completely on its head. A new element is not considered official until it's been reproduced by somebody else. Because element 118 had been predicted theoretically, Ninov expected his "discovery" to be reproduced, and then he would get the credit. But groups in Japan, Germany, and France tried to reproduce it and failed, and that left Ninov's deception naked before the world. I strongly suspect that the sorry story of the Ninov deception is only an anomaly with regard to my third condition, although future cases of scientific fraud could prove me wrong, in which case my precepts might require some modification. For the sake of both science and to a far lesser extent my own thesis, let us hope that it is a long time, if ever, before we encounter any more such cases.

Finally, we come to high-temperature superconductivity, a phenomenon that the best theoretical work had seemingly ruled out, but that turned out nevertheless to be real. It's living proof that nature likely has endless surprises for us and a cautionary reminder that we shouldn't always pay too much attention to the received wisdom. Thus far, however, its promise of commercial applications has proved to be largely ephemeral. Determining how high-temperature superconductivity actually works and whether its applications may one day offer useful solutions to real-world problems is a wonderful scientific challenge. It is one that lies beyond the scope of this volume, but I look forward to seeing it addressed in the laboratory and perhaps one day in a future book.

Here's another item for a future book: Scientists sometimes take risks. The thing about risks is just that; they're risky. Sometimes they turn out well and sometimes not. Thus, for example, Fleischmann and Pons took a risk when they announced the discovery of cold fusion, and that risk turned out badly for them. On the other hand, Bednorz and Müller took a big risk when they hid from their employer the fact that they were working on high-temperature superconductivity, and that risk turned out supremely well for them, Nobel Prizes and all.

One other thing we have learned is that Karl Popper's dictum, that science proceeds by finding things wrong, is itself questionable because the Reward System in science is set up principally to reward people for being right. An anecdote from my own experience will help to illustrate the point.

When, in 1967, I was a young postdoc at the University of Rome (*La Sapienza*) I spent the first half of the year casting about without success, looking for an experiment I could do. Then I had an idea: The work of the laboratory I was in had led to a new theory called the Huang-Olinto theory, after Kerson Huang, a professor at MIT, and his student S. A. Olinto. What I thought of was a clever and subtle way to test the theory to see if it was right. I enlisted the help of two young researchers in the group, Umberto Buontempo and Massimo Cerdonio, and the three of us set out to do my experiment.

When the experiment was done, it was clear that the theory was wrong. We had, with perfect Popperian rigor (although I doubt that at that point in my career I had paid much attention to Karl Popper), taken a very good theory and proved it wrong so that it had to be replaced by something better. If the theory had been right, our experiment would have proved it was right, and Huang and Olinto together with Buontempo, Cerdonio. and Goodstein would today be famous in the world of physics, but the theory turned out to be wrong. I wrote to Kerson Huang and told him the result of our experiment, and although he protested at first, he soon came around to admitting that our experiment was correct and his theory was wrong.

The net result was that both the theory and the experiment that proved it wrong were quickly and quietly forgotten in the world of physics. That was our reward for being good Popperians. It was a lesson to remember. It's far better to prove a theory is right than to prove it is wrong.

One take-home lesson in all of this is that scientific fraud consists of an explicit and well-defined act: faking or fabricating data or plagiarism. It is, of course, possible to violate the high standards of scientific practice in many ways, such as sloppy data-handling or extrapolating results far beyond their reasonable application. Such transgressions may be unfortunate for science, but they don't amount to fraud. Fraud consists of ffp. It's as simple as that.

Nevertheless, as the cases we've looked at should have made clear, fraud is not as easy to identify as one might suppose. Each case is unique, and must be carefully considered to see if fraud actually occurred. Thus in the case of Robert Millikan, it was a close call, but we decided that fraud had not been committed. But in the case of Vipin Kumar fraud had been committed and it was duly detected and punished. By looking at a few real-life examples we've seen that it was possible to determine that fraud was committed in one case and not in another. Those conclusions came from a careful consideration of the details of each case.

⌐ I hope that the reader will close this book with a deeper appreciation of how science (and scientists) actually work. If so, you will have an understanding grounded in the reality, not theory, of what science is. You will be able to apply the principles described in this book in looking at future cases, and of course, avoid committing fraud yourself.

⌐ Appendix
Caltech Policy on Research Misconduct

Note: This policy is intended to be compatible with the government-wide policy of the United States. It is referred to as a policy on research misconduct rather than research fraud because the government did not want the Institution responsible for proving that fraud had occurred.

Preamble

Research misconduct is historically a rare occurrence, especially at Caltech, where all members of the community are bound by a very effective code of honor. However, should an instance arise of either real or apparent misconduct, the Institute must act swiftly and decisively, while affording maximum possible protection both to the "whistle blower" (complainant) and to the accused (respondent). That is the intent of this policy.

The term *research misconduct* has been chosen instead of the narrower *scientific misconduct* to describe this policy. It refers to all research conducted at the Institute. The chair of each division is responsible for informing the division's faculty, staff, and students of the Institute's policy with regard to research misconduct, and for interpreting this policy. This policy is not intended to deal with other problems, such as disputes over order of authorship, or violation of Institute or federal regulations, that do not amount to research misconduct.

Definitions

Research misconduct is defined as fabrication, falsification, or plagiarism in proposing, performing, or reviewing research, or in reporting research results.

- Fabrication is making up data or results and recording or reporting them.
- Falsification is manipulating research materials, equipment, or processes, or changing or omitting data or results such that the research is not accurately represented in the research record.
- Plagiarism is the appropriation of another person's ideas, processes, results, or words without giving appropriate credit.
- Research misconduct does not include honest error or differences of opinion.

Findings

A finding of research misconduct requires that:

- There be significant departure from accepted practices of the scientific community for maintaining the integrity of the research record;
- The misconduct be committed intentionally, or knowingly, or in reckless disregard of accepted practices; and
- The allegation be proven by a preponderance of evidence.

Procedure

The procedures to be followed have three stages: Inquiry, Investigation, and Adjudication, or Resolution. These are the stages required by regulations issued by the federal government applicable to sponsored research. Those responsible for conducting each phase should bear in mind the following important responsibilities:

1. The institute must vigorously pursue and resolve all charges of research misconduct.
2. All parties must be treated with justice and fairness, bearing in mind the vulnerabilities of their positions and the sensitive nature of academic reputations.
3. Confidentiality should be maintained to the maximum practical extent particularly in the inquiry phase.
4. All semblance of conflict of interest must rigorously be avoided at all stages.
5. All stages of the procedure should be fully documented.
6. All parties are responsible for acting in such a way as to avoid unnecessary damage to the general enterprise of academic research. Nevertheless, the institute must inform appropriate government agencies of its actions, and if it is found that misleading data or information have been published, the institute is responsible for setting the public record straight, for example, by informing the editors of scholarly or scientific journals.

A. Inquiry

The purpose of this stage is to determine, with minimum publicity and maximum confidentiality, whether there exists a sufficiently serious problem to warrant a formal investigation. It is crucial at this stage to separate substantive issues from conflicts between colleagues that may be resolved without a formal investigation.

1. Initiating the Inquiry

All allegations of research misconduct arising from inside or outside the institute should be referred directly to the division chair (DC) concerned. If more than one division is involved, more than one DC may be informed. If either the complainant or the respondent perceives a possible conflict of interest the case may be taken directly to the provost, who will act as prescribed below for DCs, but the DC must be informed immediately and confidentially. A DC may initiate an inquiry without a specific complaint if it is felt that evidence of suspicious academic conduct exists.

When a complaint comes forth, the DC's first job is to provide confidential counsel. If the issue involved does not amount to research misconduct, satisfactory resolution through means other than this policy should be sought. However, if there is an indication that research misconduct has occurred, the DC must pursue the case even in the absence of a formal allegation. Moreover, the case must be pursued to its conclusion even if complainant(s) and/or respondent(s) resign from their positions at the institute.

The DC should also counsel those involved that, should it be found at either the inquiry or the investigation stage that the allegations were both false and malicious, confidentiality may not be further maintained and, in fact, sanctions may be brought to bear against the complainant.

2. Inquiry Procedure

The DC is responsible for conducting the inquiry (except, as noted above, where a conflict of interest might be perceived). The DC may call upon one or more senior colleagues for help where specific technical expertise is required, but this need should be carefully weighed against the importance of confidentiality at this stage. Confidentiality is likely to be a rapidly decreasing function of the number of persons involved in the inquiry.

The DC may wish to notify the president and provost, and call upon institute legal counsel at this stage. Every effort should be made to make personal legal counsel unnecessary for either complainant or respondent at this and all other stages, but all parties should recognize the institute counsel always acts on behalf of the institute, not one or the other party.

An inquiry is formally begun when the DC notifies the respondent in writing of the charges and process to follow. This and all other documents are to be preserved in a secure file in the division offices for at least three years.

The nature of the inquiry will depend on the details of the case, and should be worked out by the DC in consultation with the complainant and respondent, with any colleague the DC calls on for assistance, and with institute legal counsel. At this stage, every effort should be made to keep open the possibility of resolving the issue without damage to the position or reputation of either the complainant or the respondent. However, the DCs primary allegiance is not to the individuals but to the integrity of academic research, and to the institute. If research misconduct has been committed, it must not be covered up.

The inquiry should be completed and a written record of findings should be prepared, within 30 days of its initiation.

If the 30-day deadline cannot be met, a report should be filed citing progress to date and the reasons for the delay, and the respondent and other involved individuals should be informed.

3. Findings of the Inquiry

The inquiry is completed when a judgment is made of whether a formal investigation is warranted. An investigation is warranted if a reasonable possibility of research misconduct exists. A written report shall be prepared that states what evidence was reviewed, summarizes relevant interviews, and includes the conclusions of the inquiry. The individual(s) against whom the allegation was made shall be given a copy of the report of the inquiry. If they comment on that report, their comments may be made part of the record. The DC must inform the complainant whether the allegations will be subject to a formal investigation.

If the allegation is found to be unsupported but has been made in good faith, no further action is required, aside from informing all parties, and attempting to heal whatever wounds have been inflicted. If confidentiality has been breached, the DC may wish to take reasonable steps to minimize the damage done by inaccurate reports. If the allegation is found not to have been made in good faith, the DC should inform the provost and the president, who will consider possible disciplinary action.

If a complainant is not satisfied with a DC's finding that the allegations are unsupported, the result may be appealed to the provost, or if the provost has made the finding, to the president.

4. Notifications

The relevant responsible agency (or agencies in some cases) should be informed of the allegation upon completion of an inquiry, if (1) the allegation involves federally funded research (or an application for federal funding) and meets the federal definition of research misconduct, which is the same as the one given above, and (2) there is sufficient evidence to proceed to an investigation. The relevant responsible agency should continue to be informed of the progress of the investigation, its outcome, and any actions taken.

- Other Reasons to Notify the Agency. At any time during an inquiry or investigation, the institute will notify the relevant federal agency if public health or safety is at risk; if agency resources or interests are threatened; if research activities should be suspended; if there is reasonable indication of possible violations of civil or criminal law; if federal action is required to protect the interests of those involved in the investigation; if the provost and DC believe the inquiry or investigation may be made public prematurely so that appropriate steps can be taken to safeguard evidence and protect the rights of those involved; or if the scientific community or public should be informed.

B. Investigation

An investigation is initiated within 30 calendar days when an inquiry results in a finding that an investigation is warranted. The purpose of the investigation is to determine whether research misconduct has been committed. If an investigation is initiated, the provost and DC should decide whether interim administrative action is required to protect the interests of the subjects, students, colleagues, the funding agency, or the institute while the investigation proceeds. Possible actions might include temporary suspension of the research in question, for example. If there is reasonable indication of possible criminal violations, cognizant authorities must be informed by the provost within 24 hours. Note the provisions of section A.4 above requiring the institute to notify the agency if it ascertains at any stage of the inquiry or investigation that specified conditions exist.

1. The Investigation Committee

The provost, in consultation with the DC, shall appoint an Investigation Committee. The principal criteria for membership shall be fairness and wisdom, technical competence in the field in question, and avoidance of conflict of interest. Membership of the committee need not be restricted to the faculty of the institute.

The respondent and complainant should be given an opportunity to comment, in writing, on the suitability of proposed members before the membership is decided. The committee should be provided with a budget that will enable it to perform its task. The provost and DC should write a formal charge to the committee, informing it of the details of its task.

2. The Investigation Process

Once the Investigation Committee is formed, it should undertake to inform the respondent of all allegations so that a response may be prepared. It is assumed that all parties, including the respondent, will cooperate fully with the Investigation Committee. The committee should call upon the help of institute legal counsel in working out the procedure to be followed in conducting the investigation. The complainant and respondent should be fully informed of the procedure chosen.

At this stage, the demands of confidentiality become secondary to the necessity that a vigorous investigation make a conclusive determination of the facts. Nevertheless, every attempt should be made to protect the reputations of all parties involved.

The investigation should be completed, and a full report filed with those parties requiring notice within 120 days of its initiation. If this deadline cannot be met, an interim report of the reasons for delay and progress to date should be filed with appropriate persons and agencies.

A draft of the committee report should be submitted to both complainant and respondent for comment before the final report is written. The respondent should be given the opportunity for a formal hearing before the Investigation Committee. Institute legal counsel should be called upon to assist in working out the procedure to be followed in conducting such a hearing.

If an investigation results in a finding, based on a preponderance of the evidence, that research misconduct occurred, an adjudication or resolution phase follows whereby the recommendations are reviewed and appropriate action determined.

C. Resolution

Adjudication or resolution decisions are separated organizationally from the agency's or research institution's inquiry and investigation processes. Any appeals process should likewise be separated organizationally from the inquiry and investigation.

The committee findings may be grouped into two broad categories:

1. No Finding of Research Misconduct

All federal agencies or other entities initially informed of the investigation should be notified promptly. A full record of the investigation should be retained by the institute in a secure and confidential file for at least three years. The provost and DC should decide what steps need to be taken to clear the record and protect the reputations of all parties involved.

If the allegations are found to have been maliciously motivated, the provost and DC may wish to recommend to the president appropriate disciplinary action. If the allegations are found to have been made in good faith, steps should be taken to prevent retaliatory actions.

2. Finding of Research Misconduct

The provost and DC should decide on an appropriate course of action to deal with misconduct, to notify appropriate agencies, and to correct the scholarly or scientific record. The provost and DC should forward the committee report to the president with a recommendation of sanctions and other actions to be taken. Possible sanctions include:

- Removal from the project
- Letter of reprimands
- Special monitoring of future work
- Probation or suspension
- Salary or rank reduction
- Termination of employment

The president should review the full record of the inquiry and investigation. The respondent may at this stage appeal to the president on grounds of improper procedure or a capricious or arbitrary decision based on the evidence in the record. New evidence may lead the president to call for a new investigation or further investigation, but not to an immediate reversal of the finding. After hearing any appeal and reviewing the case, the president should make a decision, or, in appropriate cases, recommend a final disposition to the Board of Trustees. The decision of the board is final. In deciding what administrative actions are appropriate, the president should consider the seriousness of the misconduct, including whether the misconduct was intentional or reckless; was an isolated event or part of a pattern; had significant impact on the research record; and had significant impact on other researchers or institutions.

For research sponsored by a relevant responsible agency (or agencies) a final report should be submitted to describe the policies and procedures under which the investigation was conducted, how and from whom information was obtained relevant to the investigation, the findings, and the basis for the findings, and include the actual text or an accurate summary of the views of any individual(s) found to have engaged in misconduct, as well as a description of any sanctions or other administrative action taken by the institute.

In addition to regulatory authorities and sponsors, all interested parties should be notified of the final disposition of the case and provided with any legally required documentation. The list may include:

- The complainant
- Coauthors, coinvestigators, collaborators
- Editors of journals that have published compromised results
- Professional licensing boards and professional societies
- Other institutions that might consider employing the respondent
- Criminal authorities

⌐ Acknowledgments

I am happy to acknowledge the many contributions of Heidi Aspaturian to this volume, including editing the book and providing its initial title, and for thoughtful suggestions concerning its shape and substance. I would also like to thank Sara Lippincott and Judy Goodstein, both of whom contributed substantially to the way the book came out. I had many fruitful conversations with my colleague Jim Woodward, who taught me that philosophy is a serious matter. Jim Strauss was most helpful in providing the details of his investigations of Vipin Kumar and James Urban. I would like to acknowledge also the three Caltech provosts under whom I served—Barclay Kamb, Paul Jennings, and Steve Koonin. Finally, I would like to thank Caltech General Counsel Harry Yohalem for permission to reprint the *Caltech Policy on Research Misconduct* at the end of the book.

Portions of this work appeared in different form in the journals *American Scholar, American Scientist, Physics World,* and *Accountability in Research* and in *Engineering & Science,* the research magazine of the California Institute of Technology.

⌐ Notes

One
Setting the Stage

1. Richard P. Feynman, *Surely You're Joking, Mr. Feynman!* (New York: W.W. Norton, 1985), p. 343: "I'm talking about a specific, extra type of integrity that is not lying, but bending over backwards to show how you're maybe wrong, that you ought to have when acting as a scientist. And this is our responsibility as scientists, certainly to other scientists, and I think to laymen."

2. Peter Urbach, *Francis Bacon's Philosophy of Science: An Account and a Reappraisal* (La Salle, IL: Open Court, 1987). Urbach argues that Bacon's views about method were actually much more sophisticated.

3. Karl Popper, *The Logic of Scientific Discovery* (New York: Routledge, 2nd edition, 2002).

4. Feynman, *Surely You're Joking*, p. 341.

5. U.S. Dept. of Health and Human Services, Public Health Service, "Responsibilities of Awardee and Applicant Institutions for Dealing with and Reporting Possible Misconduct in Science," *Federal Register* 54:32446-51 (1989).

6. Jonathan R. Cole and Stephen Cole, "The Ortega Hypothesis," *Science* 178:4059, 368–75 (1972).

7. Robert K. Merton, "The Matthew Effect in Science," *Science* 159:3810, 56–63 (1968).

Two

In the Matter of Robert Andrews Millikan

1. For a biography of Millikan, see *The Rise of Robert Millikan*, by Robert H. Kargon (Ithaca, NY: Cornell University Press, 1982).
2. The discovery of the electron is described in J. J. Thomson, *The Corpuscular Theory of Matter (1907)* (Whitefish, MT: Kessinger Publishing, 2008).
3. Earnest C. Watson, unpublished lecture, remarks at the dedication of Millikan Laboratory, Pomona College. Caltech Archives, Watson papers, box 3.12.
4. Harvey Fletcher, "My Work with Millikan on the Oil-Drop Experiment," *Physics Today*, June 1982, p. 43.
5. C. Ian Jackson, *Honor in Science* (Sigma Xi, 1984), p.12. The new version is *The Responsible Researcher: Paths and Pitfalls*, by John F. Ahearn (Sigma Xi, 1999).
6. Charles Babbage, *Reflections on the Decline of Science in England and on Some of Its Causes* (Whitefish, MT: Kessinger Publishing, 2004).
7. R. A. Millikan, "On the Elementary Electrical Charge and the Avogadro Constant," *Phys. Rev.* 2:2, 109–43 (1913).
8. Caltech Archives, "Oil-Drop Experiment," 1911–12, Robert Andrews Millikan Collection, boxes 3.3 and 3.4.
9. Allan Franklin, "Millikan's Published and Unpublished Data on Oil Drops," *Hist. Studies in the Phys. Sci.* 11, 185–201 (1981).
10. Gerald Holton, "Subelectrons, Presuppositions, and the Millikan-Ehrenhaft Dispute," *Hist. Studies in the Phys. Sci.* 9, 166–224 (1978).
11. Millikan, *The Electron: Its Isolation and Measurements and the Determination of Some of Its Properties* (Chicago: University of Chicago Press, 1917).
12. Margaret W. Rossiter, *Women Scientists in America: Struggles and Strategies to 1940* (Baltimore: Johns Hopkins University Press, 1982), p. 192.

13. Judith R. Goodstein, *Millikan's School* (New York: W.W. Norton, 1991), p. 97.

Chapter Three
Bad News in Biology

1. For an overview of this case, see Leslie Roberts, "Misconduct: Caltech's Trial by Fire," *Science* 253:5026, 1344–47 (1991). Reprinted in Deni Elliott and Judy E. Stern, eds., *Research Ethics: A Reader* (Hanover, NH: Institute for the Study of Applied and Professional Ethics at Dartmouth College, 1997).
2. Roberts, "Misconduct: Caltech's Trial by Fire," p. 1344.
3. Ibid., p. 1345.
4. Vipin Kumar, letter to *Science*, 254:5035, 1090 (1991).
5. Eli Sercarz, letter to *Science*, 254:5035, 1090 (1991).
6. Office of Research Integrity, Annual Report, 1996 Highlights, p. 47.
7. J. L. Urban, S. J. Horvath, and L. Hood, *Cell* 59: 257–71 (1989).

Chapter Four
Codifying Misconduct: Evolving Approaches in the 1990s

1. Office of Inspector General, Semiannual Report to the Congress, No. 2, October 1, 1989–March 31, 1990, NSF, p. 20.
2. Suzanne Hadley, Muriel Jeannette Whitehill Lecture Series in Biomedical Ethics at UC, San Diego, October 17, 1991.
3. Sources that provide a range of perspectives on the Baltimore and Gallo cases include Daniel J. Kevles, *The Baltimore Case: A Trial of Politics, Science, and Character* (New York: W.W. Norton, 1998); Horace Freeland Judson, *The Great Betrayal: Fraud in Science* (San Diego, CA: Harcourt, 2004); Robert C. Gallo, *Virus Hunting: AIDS, Cancer, and the Human Retrovirus:*

A Story of Scientific Discovery (New York: Basic Books, 1991); and John Crewdson, *Science Fictions: A Scientific Mystery, a Massive Cover-Up and the Dark Legacy of Robert Gallo* (Boston: Little, Brown, 2002).

4. David Weaver et al., "Altered Repertoire of Endogenous Immunoglobulin Gene Expression in Transgenic Mice Containing a Rearranged Mu Heavy Chain Gene," *Cell* 45:2, 247–59 (1986).

5. Draft report, "Comprehensive Review of Dr. Robert Gallo's 1983–84 HIV Research, OSI 89-67," 1991.

6. "Misconduct in Science and Engineering Research: Final Rule," *Federal Register* 56:93, 22287–90 (1991).

7. "Federal Research Misconduct Policy," *Federal Register* 65:235, 76260-4 (2000).

8. Ibid.

Chapter Five
The Cold Fusion Chronicles

1. Irving Langmuir, "Pathological Science," *Physics Today*, vol. 42, October, pp. 36–48 (1989).

2. Martin Fleischmann and Stanley Pons, "Electrochemically Induced Nuclear Fusion of Deuterium," *Jour. Electroanalytical Chem.* 261:2, 301–8 (1989).

3. See, for example, Christopher Joyce, "Physics Community Strikes Back in Debate over Cold Fusion," *New Scientist*, May 6, 1989; Malcolm W. Browne, "Physicists Debunk Claim of a New Kind of Fusion," *New York Times*, May 3, 1989.

4. Marlise Simons, "Italian Researchers Report Achieving Nuclear Fusion," *New York Times*, Apr. 19, 1989.

5. J. G. Bednorz and K. A. Müller, "Possible High T_c Superconductivity in the Ba–La–Cu–O System," *Zeit. für Physik B* 64:2, 189–93 (1986).

6. Rudolf L. Mössbauer, "Kernresonanzfluoreszenz von Gamma-strahlung in Ir[191]," *Zeit. für Physik A Hadrons and Nuclei* 151:2, 124–43 (1958).

Chapter Six
Fraud in Physics

1. Jan Hendrik Schön et al., "Field-Effect Modulation of the Conductance of Single Molecules," *Science* 294:5549, 2138–40 (2001).
2. Malcolm R. Beasley et al., "Report of the Investigation Committee on the Possibility of Scientific Misconduct in the Work of Hendrik Schön and Co-authors," September 2002, Bell Labs.
3. V. Ninov et al., "Observation of Superheavy Nuclei Produced in the Reaction of [86]Kr with 208Pb," *Phys. Rev. Lett.* 83:6, 1104–7 (1999).
4. Rochus Vogt et al., "Report of the Committee on the Formal Investigation of Alleged Scientific Misconduct by LBNL Staff Scientist Dr. Victor Ninov," March 27, 2002.
5. Bertram Schwarzschild, "Lawrence Berkeley Lab Concludes that Evidence of Element 118 was a Fabrication," *Physics Today*, 55:9, 15–17 (2002).
6. Charles Seife, "Heavy-Element Fizzle Laid to Falsified Data," *Science* 297:5580, 313–15 (2002).

Chapter Seven
The Breakthrough That *Wasn't* Too Good to Be True

1. J. G. Bednorz and K. A. Müller, "Possible High T_c Superconductivity in the Ba–La–Cu–O System," *Zeit. fur Physik B* 64:2, 189–93 (1986).

2. James Gleick, "Discoveries Bring a 'Woodstock for Physics,'" *New York Times*, March 20, 1987.

3. Heike Kamerlingh-Onnes, "The Superconductivity of Mercury," *Comm. Phys. Lab. Univ. Leiden*, nos. 122 and 124 (1911).

4. M. K. Wu et al., "Superconductivity at 93K in a New Mixed-Phase Y-Ba-Cu-O Compound System at Ambient Pressure," *Phys. Rev. Lett.*, 58:9, 908–10 (1987).

5. See, for example, P. Dai et al., "Synthesis and Neutron Powder Diffraction Study of the Superconductor $HgBa_2Ca_2Cu_3O_{8+\delta}$ by Tl Substitution," *Physica C:Superconductivity* 243:3–4, 201–6 (1995) doi:10.1016/0921-4534(94)02461-8, and L. Gao et al., "Superconductivity up to 164 K in $HgBa_2Ca_{m-1}Cu_mO_{2m+2+\delta}$ (m = 1, 2, and 3) under Quasihydrostatic Pressures," *Phys. Rev. B* 50:6, 4260–63 (1994) doi: 10.1103/PhysRevB.50.4260.

Index

Breuning, Stephen, 2
Brigham Young University, 70, 76
Broad, William, 43–44
Bronze Age, 123
Brownian motion, 29, 32, 40, 44
bullet trains, 120, 124–25
Buontempo, Umberto, 131–32
Burt, Cyril, 2

California Institute of Technology
(Caltech), xiii, 11, 82; as all-male
school, 50; Baltimore and, 63;
Barnes and, 73–74, 79, 87, 90,
94; Biology Division of, 51–57;
cold fusion and, 73–74; formal
misconduct policy for, 52, 65–68,
135–45; Hood and, 51–55; Jews in,
50; Koonin and, 73–74, 87, 89, 94;
Kumar and, 51–57, 128, 132; Lewis
and, 73–74, 79, 87, 89, 92, 94;
Millikan and, 48–50; Office of Research Integrity (ORI) and, 54–55,
66; Rossman and, 116; Scientific
Ethics course at, xi–xii, 94, 104; six
divisions of, 65; Urban and, 51–52,
55, 57, 128; Watson Lecture series
and, 107
California Public Records Act, 104
calorimetry, 69
Caltech Archives, 34
Caltech Policy on Research Misconduct:
Board of Trustees and, 144–45;
definitions and, 136; division chair
(DC) and, 138–44; findings and,
136; Goodstein and, 52, 65–68;
inquiry and, 137–40; investigation and, 141–43; notifications
and, 140–41; preamble, 135–36;

procedure in, 137–45; resolution
and, 143–45
career pressure, 3–4, 51, 56, 101,
129–30
cathode-ray tubes, 30
Cell journal, 55–56, 61–62
ceramics, 84, 116
Cerdonio, Massimo, 131–32
chemistry, 19; Boyle and, 25; cold fusion and, 72–73, 78, 92–93; women
and, 49
China, 108
Christian Democrats, 74
Chu, Paul C. W., 108, 114–16
cloud chambers, 30–32
cold fusion, xiii, 108; American Physical Society and, 72–73; Barnes and,
73–74, 79, 87, 90, 94; calorimetry
and, 69; chemists and, 72–73, 78,
92–93; continued research on, 89;
cosmic rays and, 90; deuterium
and, 71–72, 76, 80–85, 89–94,
129; dry fusion and, 78; economic
impact of, 72; electrolysis and, 72,
76, 78, 91, 129; experiments and,
70, 86–88; falsifiability and, 79;
Fleischmann and, 69–73, 76–83,
88, 91–93, 131; Fourth International Conference on Cold Fusion and,
69; funding and, 76–78, 89; gamma
rays and, 83, 85–86; heat from, 76,
79, 82–83, 86–95; high-temperature superconductivity and, 83–84,
93; hot fusion and, 71–72, 81–82,
89; Jones and, 70, 76, 78, 80–82,
88; Koonin and, 73–74, 87, 89,
94; Lewis and, 73–74, 79, 87, 89,
92, 94; media and, 70–71, 75–79,

Office of the Inspector General
(OIG), 60, 64
Olinto, S. A., 131
Oppenheimer, J. Robert, 97
organic semiconductors, 98–99
Ortega hypothesis, 15, 15–16
Ortega y Gasset, José, 15
O'Toole, Margot, 61

palladium, 72, 76, 82, 94, 129
particle accelerators, 102–6
particle detectors, 30–32
Pasteur Institute, 62, 64
pathological science, 70
patrons, 25–26
peer review, 17, 23–24, 53, 55, 69–70,
79
Perrin, Jean, 29
personal gain, 6–8
Ph.D.s, 3, 20, 22–23, 29, 32, 51, 54, 97
photoelectric effect, 29
photons, 81, 85–86
Physical Review, The, 46
Physical Review Letters, 103
physics: atomic bombs and, 97; cold
fusion and, 69–95; element 118
and, 97, 103–4, 106, 130; low-
temperature, 107; Millikan and,
29–50; single-molecule transistor
and, 99; trained intuition and,
83–86
Physics Today journal, 106
plagiarism, 60, 64, 67, 132, 136
Planck, Max, 29
plasma, 71–72, 82–83, 89
platinum, 72, 94, 129
Policy on Research Misconduct
(Caltech), xii–xiii

Pons, Stanley: celebrity status of, 69;
cold fusion and, 69–83, 88, 91–93,
131; heat measurement of, 82;
lack of control experiments by, 93;
Lewis criticisms and, 92; Nobel
Prize and, 93–94; ordinary water in
experiments of, 91, 93; as outcast,
94; press conference of, 70–71;
reproducibility and, 73
Pool, Sandy, 52
Popovic, Mikulas, 61–62, 64
Popper, Karl, 9–11, 79, 87, 127,
131–32
prestige. *See* Reward System
Principia (Newton), 26
principle of relevance, 9
priority, 26
*Proceedings of the National Academy of
Sciences (PNAS)*, 54–55
professorships, 22
protons, 30, 72, 81, 103
psychology, 2, 61
Ptolemy, Claudius, 44
Public Health Service (PHS), 59–60

Quantum Hall effect, 98, 101
quantum physics, 29

racial issues, 29, 49–50
radioactive waste, 71
regulations. *See* government
reproducibility, 4–5, 61–62; cold
fusion and, 73, 75–76, 79, 86–87,
93–94; element 118 and, 103–4,
130; high-temperature superconductivity and, 93, 116; organic
semiconductors and, 98–99; Schön
and, 98–99, 102

research: Authority Structure and, 7, 14, 18–27, 127; *Caltech Policy on Research Misconduct* and, 65–68, 135–45; cold fusion and, 69–95; cooking and, 33, 35, 44; data fabrication and, 33 (*see also* data fabrication); doctoral degrees and, 19; element 118 and, 97, 103–4, 106, 130; first laboratories for, 25; government regulations, 59 (*see also* government); Health Research Extension Act and, 59; "ladder of fame" in, 20–22; multiple sclerosis and, 51, 55, 57; National Research Council and, 19; Office of Research Integrity (ORI) and, 54–55, 60–63, 66–67, 128; Office of Scientific Integrity (OSI) and, 59–68, 128; peer review and, 17, 23–24, 53, 55, 69–70, 79; Reward System and, 16–27, 127, 131; semiconductors and, 98–99; universities and, 19

Revolt of the Masses, The (Ortega y Gasset), 15–16

Reward System, 16; funding and, 21–22; immortality and, 18, 22; "ladder of fame" for, 20–22; letters of recommendation and, 23–24; Matthew effect and, 25; named discoveries and, 22; Nobel Prizes and, 20; Popper and, 9–11, 79, 87, 127, 131–32; resting on laurels and, 21–22; roots of, 25–26; scientific societies and, 19–27; serendipity in, 24; tenure and, 21; university presidents and, 18

Rhine, J. B., 70

Ritalin, 2

Robert Andrews Millikan Professor of Biology, 63

Rockefeller University, 62

Rome, 75–77, 90–91

Rossman, George, 116

Royal Caroline Institute, 20

Royal Society, 25, 78

Royal Swedish Academy of Sciences, 20

Rutherford, Ernest, 30

scandium, 115–16

Scaramuzzi, Francesco: apparatus of, 76; cold fusion and, 74–75, 80–82, 87–89, 94; data of, 87–88; Lewis criticisms and, 91–92; neutron detection and, 76, 81; other research of, 89–90; political integrity of, 74–75; press conference of, 77; promotion of, 77

Schön, Jan Hendrik; 106, 130; allegations against, 99–100; background of, 97; Beasley committee and, 99–102; career pressure and, 101; coworker culpability and, 100; data fabrication and, 99; field-effect doping and, 98; firing of, 100; Nobel Prize and, 97; promotions of, 99; Quantum Hall effect and, 98, 101; reproducibility and, 98–99, 102; scientific papers and, 99, 101; semiconductor research and, 98–99; single-molecule transistor and, 99; superconductivity and, 98, 101; University of Konstanz and, 97–100 science: active measures for, 2; Authority Structure and, 7, 14, 18–27, 127; collaboration and, 7, 14–15, 27, 100–101; ethical principles for, 6–7; eureka moments and, 107; faith in, 1; Matthew effect and, 25; motivations for studying, 18; mystery of, 116; nature's ingenuity and, 107; Nobel Prizes and, 20; pathological, 70; peer review and, 17, 23–24, 53, 55, 69–70, 79; political parties and, 74–75; reproducibility and, 4–5; reputation and, 7; Reward System and, 16, 18–27, 127, 131; risk and, 131; self-correction and, 2; transparency in, 7; truth and, 27; wonder of, xiii–xiv

Science journal, 16, 43, 99, 106

science wars: Authority Structure and, 7, 14, 18–27, 127; gatekeepers and, 22–24; "ladder of fame" in, 20–22; priority and, 26; Reward System and, 7, 14, 16–27, 127, 131; scorekeeping in, 24; university rivalry and, 20–24

"Scientific Ethics" course, xi–xii, 94, 104

scientific method, 1; cold fusion and, 79–83, 95; data fabrication and, 2; ethical principles for, 6–7; falsifiability and, 9–11, 64, 67, 79, 131–32; hypocrisy in, 5; hypothesis and, 6; inductivism and, 8–10; as last word, 127; mediocrity and, 15–16; observation and, 8–10; Ortega hypothesis and, 15–16; universal strategies for, 13–14; winner-take-all credit system and, 13

scientific misconduct, xi–xii, 57, 95, 129; *Caltech Policy on Research Misconduct* and, 65–68, 135–45; career ladder and, 21–22; case complexities and, 64; coworker culpability and, 100; defining, 6, 13, 60–61, 64–65, 135–36; *Federal Register* and, 64, 68; Health and Human Services appeals and, 63; Imanishi-Kari and, 61–64; Kumar and, 55; learning from, 127–33; Millikan and, 33, 35, 43; Ninov and, 104–6; Office of Research Integrity (ORI) and, 54–55, 60–63, 66–67, 128; peripheral role of, 27; plagiarism and, 60, 64, 67, 132, 136; Popovic and, 61–62, 64; Schön and, 99–102. *See also* fraud

scientific papers: Brownian motion and, 32; cold fusion and, 69–70; collaboration and, 7, 14–15, 66, 100–101; elitism and, 15–16; Fletcher and, 32–33; guest authorship and, 15; hypocrisy in, 5; Millikan and, 32–33; misrepresentation and, 16–17; multiple authors and, 7; Ortega hypothesis and, 15–16; peer review and, 17, 53, 55, 79; reputation and, 23; Schön and, 99, 101

scientific societies: cold fusion and, 72–73, Reward System and, 19–27. *See also specific groups*

Second Coming, 123–24

Secret Service, 62

semiconductors, 98–99

Sercarz, Eli, 54

sexism, 48–49